Breeding and Crop Production

Breeding and Crop Production

Editor

Dr. P.C. Bandyopadhyay

GENE TECH

2007

Gene-Tech Books

New Delhi - 110 002

ISBN 81-89729-71-3

Published by : **GENE-TECH BOOKS**
4762-63/23, Ansari Road, Darya Ganj,
NEW DELHI - 110 002
Phone: 41562849
e-mail: genetechbooks@yahoo.co.in

Printed at : **Tarun Offset Printers**
Delhi

PRINTED IN INDIA

Preface

Agriculture is the process of producing food, feed, fibre and other goods by the systematic raising of plants and animals. More that 42 percent of the world's labourers are employed in agriculture, making it by far the most common occupation. Domestication of plants is done in order to increase yield, improve disease resistance and drought tolerance, ease harvest and to improve the taste and nutritional value and many other characteristics. Centuries of careful selection and breeding have had enormous effects on the characteristics of crop plants. Plant breeders use greenhouses and other techniques to get as many as three generations of plants per year so that they can make improvements all the more quickly.

In industrialised agriculture, crop improvement has often reduced nutritional and other qualities of food plants to serve the interests of producers. After mechanical tomato-harvesters were developed in the early 1960s, agricultural scientists bred tomatoes that were harder and less nutritious. In fact, a major longitudinal study of nutrient levels in numerous vegetables showed significant declines in the last 50 years. Very recently, genetic engineering has begun to be employed in some parts of the world to speed up the selection and breeding process.

This book explores innovative methods for breeding new varieties of major crops with resistance to environmental stresses that limit crop production worldwide. It provides the basic principles and techniques of agricultural breeding and examine work that has been done in relation to improving resistance in specific important world food crops. The contents of the book will help scientists and academics to examine, modify, and improve breeding and crop production programmes.

Editor

Contents

1

Habits of Crop Production

The present human population of the world is 5.7 billion, increasing at an annual rate of 1.6 %. According to the latest estimates, the rate of population growth will decrease slowly during the next century, and the world population is expected to stabilize at about 11.5 billion shortly after the year 2100.

The rapid population growth in general and the fact that most of the population growth takes place in developing countries rises the question how these additional people can be fed. Basically, there are two different possibilities to increase food production: Enlarging crop yields on already cultured land, or expanding the area of cultured land on previously uncultured land.

Reviewing the development of crop production since 1965 reveals a massive increase. The production of rice, wheat and maize, which are the most important crops for human nutrition, doubled during the last 25 years. Higher yields were possible because of the introduction of high-yielding varieties, improved cultivation techniques, fertilizer use and chemical crop protection measures—a development which has been termed "green revolution".

The intensification of crop production improved the level of food consumption in most developing countries during the last 30 years despite a world population growth of 2.4 billion in the same period. However, the intensification also led to a number of disadvantageous side-effects like erosion, deterioration of soil quality, salinization, and others. Every year, about 1 % of the cultured area has to be replaced because of these reasons. Especially in developed countries, the use of agrochemicals is a growing public concern. The amount of crop protection products used has risen steadily during the past 30 years. Presently, more than 25 billion US$ per year are spent for herbicides, fungicides, insecticides, soil disinfectants and growth regulators.

Habits of Roots in Crop Production

The roots of plants are the absorbers of water and mineral matter from the soil. These, with the carbon dioxide taken from the air, are the materials out of which not only the food for plants but the world's food supply is manufactured by plants. Green plants, only, of all living organisms, are able to build up organic food from the relatively raw materials of the inorganic world. (Certain bacteria, e.g. nitrifying and sulphur bacteria, are also able to build up organic materials. They obtain their energy by oxidizing ammonia and hydrogen sulphide and obtain carbon from the carbon dioxide of the air without the agency of sunlight.

The organic substances produced are relatively negligible in amount.) Cultivated plants almost always have a sufficient supply of carbon dioxide for normal growth. Oxygen for respiration and light are likewise usually plentiful. The temperature of the air and soil, except in high latitudes, is seldom seriously unfavourable for plant development, and it is unusual to find the physical or chemical nature of the substratum unsuitable for plant production. But in the principal agricultural sections of the United States, especially in the more and or semiarid parts, water is the chief limiting factor.

Insufficiency of water, perhaps more often than any other environmental factor, lessens crop yield. In fact, more than half of the surface of the earth receives insufficient precipitation for the most favourable growth of crops. A method of supplementing this inadequate water supply is by the application of water by irrigation, and the normal supply may be conserved by systems of tillage and crop management which aid in getting as much of the precipitation, as possible into the soil and keeping it there until it is absorbed by plants.

The great importance of the seedling establishing contact with a water supply is shown by the vigorous growth of the rootlet which usually reaches a considerable length and often begins to branch before the leaves are unfolded. In fact, the rate of this early root growth may determine the success or failure of establishment in dry situations. In some agricultural areas, on the other hand, soil water is so abundant that most crops will not grow until the wet soil is drained, often at great trouble and expense, and a supply of fresh air for the roots is admitted.

Keeping the soil in proper physical and chemical condition as regards water content, aeration, productivity, temperature, etc., so as to promote vigorous plant growth, is the purpose of tillage even in moderately moist soil where rainfall is approximately optimum. The functioning of the part of the plant aboveground is conditioned very much by the distribution and activities of the root system.

The latter may be controlled to a considerable extent by various cultural practices. Thus, it is obvious that an exact knowledge of root development of cultivated plants, of their position, extent, and activities as absorbers of water and nutrients at various stages of growth is of paramount importance to all who are engaged in crop production. Likewise, a knowledge of modifications that are produced by variations in the soil, whether due to soil structure, to excessive water content or drought, tillage, fertilizers, or other causes, is of no less importance.

Activities of Roots

The chief activities of roots, aside from growth and anchorage of the plant, are the absorption and conduction of water and nutrients. Absorption may occur throughout the entire extent of the root but takes place most actively in the younger and usually deeper parts.

Roots of cultivated plants, contrary to current opinion, are not superficial but deeply seated. Studies on the rooting habits of crops, giving a clear understanding of their great extent, have been so recent and the old idea of roots being superficial is so commonly believed that many current textbooks still speak of them as "shallow." The old viewpoint is well stated as follows in one of our best modern works on soils.

It is well known that only the top 6 or 8 inches of the soil is suited to plant life and that the, lower part, or subsoil, plays only an indirect part in plant nutrition- We shall, therefore, confine our attention almost exclusively to the surface layer.

Absorption of Water

That crops absorb water in large quantities from the subsoil and that both quantity and quality of yield can be greatly modified by the addition of fertilizers to the subsoil have been conclusively demonstrated.

In One experiment, barley was grown in the field in large cylindrical containers 18 inches in diameter and 3.5 feet deep. The containers were filled in such a, manner that each 6-inch layer occupied the same depth in the container that it had formerly occupied in the field. Each 6-inch layer of soil, of known water content, was separated from the one above and below by thin wax seals. The seals effectually prevented the movement of the water from one soil layer to another but permitted the roots to develop in a normal manner.

When the crop was ripe, it was found that the water had been absorbed from the several levels, beginning at the surface, in the following amounts: 20, 19, 16, 16, 14, 12, and 11 per cent, respectively, based on the dry weight of the soil. Moreover, it was further ascertained, from barley grown in other containers examined at various intervals, that during the period of heading and ripening of the grain, the bulk of absorption was carried on by the younger portions of the roots in the deeper soils, the surface layers by that time being quite dry.

Similar results have been obtained with rye. During the period of ripening, maximum absorption occurred at a depth of 2.5 feet and considerable quantities of water were removed at greater depths. Winter wheat behaved in a similar manner, absorbing most vigorously late in life from the deeper soil layers (2.5 feet or deeper) and reducing the water content to a much, greater degree than in the moisture layers above. These results show clearly that absorption by a plant during different stages of its life does not depend upon total root surface or root mass but is determined mainly by the area of the functioning parts of the root system.

Potatoes absorbed water to depths of 2.5 feet. Corn, which uses water extravagantly, absorbed large quantities from the third and fourth foot and smaller amounts from the fifth when grown in containers 3 feet in diameter and 5 feet deep.

Absorption of Nutrients

In other experiments, nitrates were placed in the soil in measured amounts (400 parts per million of soil) at various levels. At the time of blossoming, barley had removed 286 parts per million of the nitrates from the third 6-inch level and 135 and 168 parts per million from the next deeper layers, respectively.

At maturity, it had absorbed 186 parts per million of nitrates from the 2- to 2.5-foot layer of soil. Potato roots absorbed the nitrates at similar levels, although in slightly smaller amounts, and corn removed 203, 140, and 118 parts per million, respectively, from the third, fourth, and fifth foot of soil. In every case where roots came in contact with a fertilized soil layer, they developed much more abundantly and branched more profusely. Such a layer also retarded normal penetration of the root system into the soil below.

Other experiments in which the soil layers were impregnated with monocalcium phosphate showed similar root activities regards absorption at the several depths. These experiments show that the roots of crops were active in the absorption of both water and nutrients even at the maximum depth of their penetration. The materials necessary for food manufacture were taken from the deeper soil in considerable quantities, although to a lesser extent than from the soil nearer the surface, which the roots occupy first and, consequently, at least in annual crops, where they absorb for the longest time. It is important to note that these plants received their supply of water and nutrients from the deeper soil layers during the later stages of their development which, in some respects, are the more critical ones.

To determine the effect of the depth of fertilizers upon the yield, two series of containers were used in which barley was grown. In the first, the soil was fertilized with nitrates or phosphates at different sealed 6- inch levels. In the second series, the surface foot was fertilized and, in addition, various 6-inch levels of the subsoil. Although the plants used the largest amounts of salts from the surface foot, they also took large additional quantities from the deeper levels when they were available.

The absorption of nutrients at levels below the surface affected materially both quantity and protein content of the yield. This effect was pronounced even when the surface foot of soil was abundantly supplied with a similar nutrient. For example, sodium nitrate when applied to the surface foot increased the total dry weight by promoting heavy tillering and also added to the nitrogen content of the grain. But when available at lower levels as well as in the surface foot, it increased still further both the dry weight and the quality of the grain.

Such experiments show the importance of the subsoil as a probable source of nutrients for crops. It seems clear that an attempt should be made to promote, by cultural practices, an ample distribution of the deeper portion of the root system. In the agricultural practices of humid regions, the "rawness" or unproductiveness of subsoil has long been recognized, but it is well known and generally accepted that subsoils of arid regions are not unproductive and that, in them, plants make a good growth.

In the semiarid soils of the eastern half of Nebraska, eight successive crops of inoculated alfalfa gave almost as heavy yields on the subsoil as on the corresponding surface soils, although corn and non-inoculated legumes did poorly. That roots absorb phosphorus and potassium as well as other elements from the subsoil is indicated by the fact that plants, when potted in subsoil, will grow if nitrogen is added. In some soils, nitrogen has been shown to leach in rather large amounts to depths of 3 to 6 feet or more.

That many native species must absorb their nutrients from the subsoil, i.e., below a depth of 2 to 3 feet, is clearly indicated by the root habit. Little or no branching occurs in the surface layers, while it seems certain that morphological changes in the roots occupying the upper soil are such as to preclude absorption. Differences in the root habits of various cultivated plants may also lead to marked variations in the amount of nutrients removed from the different soil levels.

Consequently, knowledge of the extent and distribution of the roots of cultivated plants is of great practical interest. The fact that roots may absorb nutrients at deep levels in the subsoil as well as from the surface layer should be given greater attention by all plant growers. The too prevalent idea that it is mainly the surface layer of soil that supplies the plant with nutrients and that the subsoil is the crop's reservoir for water should give way to the fact that it is the whole soil mass permeated by roots that determines root activity.

Response of Roots to Environmental Factors

The habits of roots, as those of shoots, are more or less characteristic for every kind of plant. They are governed, first of all, by the hereditary growth characters of the species or variety in question. It has been shown that inbred strains of corn, for example, differ greatly in the character and extent of their root systems. "Certain strains... have such a limited and inefficient root system that they are unable to function normally during the hot days of July and August, when the soil moisture is low."

Other strains have fewer roots and a lower ratio of tops to roots. Experimental evidence has recently been found which supports the suggestion that selective absorption by individual corn plants may prove to be a very important heritable character.

A study of the inheritance of root forms in mangels and sugar beets has shown that, in general, the roots of the F_1 generation were intermediate between the parental forms. Sugar beet crosses in which wedge-shaped forms were involved proved to be exceptions. Wedge shape was completely dominant over walnut form and also over long, somewhat slender roots. Recent experiments with peas, where a dwarf variety with a short root system was crossed with a tall variety with a deep root system, indicate that root characters are hereditary and segregate out in the F_2 generation according to the Mendelian ratios.

Within the species or variety, root modifications are usually brought about by the operation of such factors as water content, aeration, soil structure, and nutrients. In fact, the character of the root system is usually an indicator of soil conditions. Of 28 native grasses and other herbs studied in two or more widely separated habitats, showed very striking changes in their root habits as to depth of penetration, and position and number of branches; one exhibited only moderate differences, and two showed practically no change.

Several shallow-rooted forest trees as well as certain deeply rooted ones do not adapt themselves to changed soil conditions but others belonging to each class show considerable plasticity. Great variability occurs in the rooting habits of fruit trees. For example, the wide adaptation of black walnut to so many soils that it is almost universally used as stock for the English walnut in California is well known.

Other fruit trees show much less plasticity, certain varieties failing unless grafted onto other stock, the roots of which adapt themselves to soils underlaid with alkali or containing excessive moisture, or to very exposed or dry situations. Sometimes, the root variation is so great and the growth habit so profoundly changed that the roots are scarcely recognizable as belonging to the same species.

Relation between Root and Soil Moisture

In studying the moisture relations of a soil, the root extent should furnish the criterion as to the depth to which soil moisture should be studied and also the maximum depth to which samples should be taken. The time, method, and amount of the application of irrigation water should be worked out in connection with their effect upon root distribution. It should also be kept clearly in mind that the ideal root system is not necessarily one with the most extensive branching but one that fully occupies the soil to an adequate depth and throughout a radius sufficient to secure enough water and nutrients at all times.

Keeping the surface soil too moist during the early life of the plant may promote a more shallow rooting habit, and the crop may later suffer from drought, unless watered very frequently. One of the most difficult problems of irrigation is to apply the water in such a way that plants are not made surface feeders. Otherwise, the natural advantages of the roots widely penetrating the subsoil for nutrients are lost. Conversely, delay in time or amount of water used may tend to promote a deeper rooting habit.

The proportion of roots to tops may be definitely increased by lowering the soil moisture. Roots of plants that mature a crop in fairly dry soil must penetrate deeply and spread widely, a distribution hindered by a very moist surface soil early in the life of the plant. Crops respond to differences in water content and aeration, both in amount and direction of growth. By varying these factors by the application of more or less water, not only the root system but also the aboveground plant parts and yields may be varied, since a close correlation exists between the growth of roots and tops.

Two-year-old alfalfa plants grown under irrigation in dry upland soil in New Mexico had roots 3 to 4 feet deep where 2 inches of water were applied at each irrigation, but they were 4.5 to 5 feet deep where 5 inches of water were applied each time. While too much water may produce yellow, shrunken kernels of wheat, it has been shown that cereals are able to utilize water up to the date of full maturity. Too little water, even after the spikelets are losing their cooler, results in checking the deposit of dry matter in the grain, and a deficiency earlier in the development of the kernel probably determines its size, even before the rate of the deposit of dry matter is checked.

Clearly, the necessary water can be applied more effectively if a knowledge of the extent and position of the root system as modified by the chemical and physical nature of the soil is known. Decreased yields may often be correlated directly with conditions influencing the development of roots. This is especially true in irrigated districts where water supply is the great limiting factor in crop production. In fact, root development often explains the reasons for differences in crop yields that are otherwise obscure.

Experiments have shown that when water is applied at the proper time, two or three irrigations give as good results as a greater number. In some areas, a method of growing wheat with a single irrigation has been worked out under which, the yields are often higher and the harvest is earlier than under the old practice of irrigating six times. Thus, there is a great saving not only of water for use on other arable lands but also of time and labour.

Too much water is frequently injurious to the soil by leaching out nutrients and in other ways. It often delays the maturing of the crop which is, consequently, more liable to rust and attack by other diseases. Frequently, both quality and quantity of yield are decreased. Since crops differ greatly in the amount of water they can profitably use, as well as in their response to adverse aerial environments, a thorough knowledge of the root systems is not only warranted but imperative for an adequate explanation of crop behaviour.

Raising the water table even temporarily by irrigation causes the death of the deeper roots in many plants and usually results in a decreased yield. The roots of some species succumb more readily than others. Among many plants, top development depends upon a sufficient root supply, as was clearly illustrated in the case of the cotton plant. The amount of shedding of leaves and bolls was directly proportional to the extent of the root system which was submerged (and died) as a result of a rising water table. But when the water table was again lowered, a new growth of tops took place simultaneously with a new growth of roots into the area thus provided for root extension.

The general shape of the root system of trees and other plants may be controlled more or less by regulating, under irrigation, the depth of the water table. If the subsoil is water-logged and thus unaerated, deeper roots will not develop or, if already grown, will soon die as the water table rises. In either case, there is a marked tendency towards the production of an abundance of roots so superficially placed that cultivation results in more or less serious root pruning. Moreover, under such conditions, plants are more sensitive to drought, temperature changes, etc. They require heavier irrigation and greater amounts of fertilizers than those more deeply rooted.

Effects of Drainage

The proper drainage of swamps and bog lands for cultivated crops should be determined with reference to root relations. Extensive experiments have shown that the water table, if at a shallow depth, determines the limit of root penetration and, to a large extent, the yield of many common meadow grasses such as timothy, meadow fescue, and bluegrass. Even when the water table is high, only rarely does a root penetrate into the saturated soil, although, in well-drained soils, these grasses are quite deeply rooted.

In pasture mixtures, this water relation may be an important factor in determining which species will thrive and become dominant and which will disappear. Many coarse marsh grasses and grass-like plants can thrive in wet situations, since their roots are built anatomically to permit of rapid gaseous exchange. But most cultivated plants require well-aerated soil. For example, corn in well-drained soil frequently penetrates 6 feet or more deep, but in peat marshes, where a system of underdrainage kept the water table almost stationary at about 2.5 feet, it, has been shown that the roots, upon reaching a level 18 inches above the water table, turned aside and failed to penetrate deeper.

This inhibition to deep root penetration clearly reflected itself in the dwarfed stature and reduced yields of the aboveground parts. Although this may be found to be an exceptional case, it clearly shows the important relation between root habit and the plan of the drainage system by which the water level in such marshes should be controlled. The maximum depth to which the water table should be lowered depends largely upon the nature of the soil as, well as upon the root habits of the crops to be grown. If the soil is coarse and capillary action consequently low, too great lowering of the water table may result in a soil too dry to afford maximum yields

Low Water Content of Soil

A relatively low water content of soil, within certain limits, stimulates the roots to greater development, resulting in a greatly increased absorbing surface. Corn grown for 5 weeks in a moist, rich, loose soil (available water content 19 per cent) had a total root area which was 1.2 times greater than that of the transpiring surface of stems and leaves. Corn, with similar hereditary characters, grown in like soil, with an available water content of only 9 per cent, had a root area 2.1 times greater than that of the top.

Similar results were obtained with 2-months-old alfalfa, although, here, the area of the taproot system was exceeded by that of the tops. Plants grown in rich silt loam soil with an available water content of 22 per cent developed a root area 66 per cent as great as the area of the tops. But in a similar soil with 10 per cent available water, the root system had 83 per cent as much area as the aboveground parts. Thus a low water content, within certain limits, stimulates increased root development, which results in a greatly increased absorbing area.

When a crop is planted too thickly, even under otherwise favourable conditions for growth, the plants do not reach maximum development, because there is not enough light, water, and nutrients for all. Under these conditions of competition, the root system has been found to be more extensive in proportion to tops than those of crops less closely spaced. Where decrease in light is the most important factor, the growth of the shoot is made more and more at the expense of the dry weight of the root.

But where the soil is very dry, root development is greatly retarded or even ceases, and the aboveground parts are dwarfed accordingly. For example, on the short-grass plains, roots of alfalfa and wheat, which normally penetrate several feet deep, although more profusely branched, are almost entirely confined to the surface 2 feet of soil because of lack of sufficient water to promote growth in the subsoil. Why some crops are better adapted to semiarid regions than others may sometimes be explained, at least in part, by a study of the root habit, although the ability of a plant like sorghum to endure drought by remaining relatively quiescent is an exceedingly important character.

Sorghums when compared with corn were found to have better developed root systems as regards degree of branching in relation to the extent of tops. These studies were carried on in rather lightly irrigated sandy loam soil in southwestern Kansas. Dwarf milo, Blackhull kafir, and a dent corn were grown in alternate rows. The leaf area of the corn at all stages of its growth was approximately twice as great as that of the Dwarf milo and at least one and a half times that of Blackhull kafir. Both sorghums had, in all stages of their development, a main root system just as extensive as that of the corn and, in addition, possessed twice as many secondary roots.

Effects of Fertilizers

A knowledge of root systems is fundamental in the proper application of fertilizers. In fact, it should be an important basis for determining not only the time but also the manner and depth of application. Likewise in investigations of soil productiveness, root extent and activity should determine the portions of the soil and subsoil to be studied.

Crops grown in rich soil have roots that are shorter, more branched, and more compact than those grown in similar but poorer soil. Sachs, over half a century ago, demonstrated that the more concentrated the nutrient solution, the shorter the roots, and Liebig stated that plants search for food as if they had eyes. It has been known for a long time that plants grown in soils with alternate layers enriched with nutrient solution branch much more profusely in these layers. Experiments so arranged that one-half of the root system of peas, for example, grew in soil rich in nitrates and the other half in poor soil gave similar results.

Of course, enriched soil promotes better shoot development which, in turn, furnishes a greater food supply for further root growth. The correlation of the growth of roots and tops is usually pronounced. Injury to one also hinders the growth of the other.

When roots enter a soil area enriched by the decay of former roots, the greater degree of branching is often very marked. They frequently follow the path of their predecessors for considerable distances, branching in great profusion. Similar branching, which may be due partly to better aeration, frequently occurs in earthworm burrows.

Marked contrasts in the degree of ramification of roots as they penetrate different soil strata are often to be attributed to differences in richness of soil. For example, sugar beets grown in fine sandy loam stratified at various depths with layers of clay have been found to branch much more profusely in these clay layers richer in nutrients than in the soil above or below. Frequently, root stratification is due to soil moisture, but the factors of water content and nutrients often operate together. In tree plantations of green ash on the short-grass plains, the surface 6 inches of soil is literally filled with great masses of finely branched tree rootlets. Under the light precipitation, most of the absorption must take place from this humus-enriched surface soil. However, many large but very poorly branched roots penetrate to a depth of 8 feet or more into the dry subsoil.

Similar root layering has been found among many varieties of grapes. While most of the laterals of 6-year-old vines were present in the surface foot of soil, ranging from 10 to 20 feet from the base of the plant, many penetrated several feet deep and occasionally a root reached a depth of over 7 feet. In dune soils where the organic matter is mainly responsible for the water capacity, the roots of many plants tend to occupy the upper layers where the humus and moisture are most abundant. It has been shown experimentally that in every case where roots came in contact with a soil layer rich in nitrates, they not only developed much more abundantly and branched more profusely but failed to penetrate as far into the deeper soil.

On the other hand, it has been shown that wheat and barley seedlings grown in both soil and culture solutions low in nitrates produced remarkably extensive root systems, although the shoots were small. Similar results, but to a less marked degree, were obtained when potassium salts were deficient. Thus, it seems clear that the depth at which the fertilizer is placed in field practice considerably affects root penetration and development.

Fertilizing the surface layers of soil, especially with nitrates, and thus stimulating surface root production in regions where these layers have very little or no available water during periods of drought appears to be distinctly detrimental to normal crop production. The effect of phosphates in promoting root growth in length and number of branches has long been recognized in agricultural practice. Dressings of phosphates are particularly valuable whenever greater root development is required than the soil conditions normally bring about... Phosphates are needed also for shallow-rooted crops with a short period of growth... Further, they are beneficial wherever drought is likely to set in because they induce the young roots to penetrate rapidly into the moisture layers of the soil below the surface.

Wheat on land treated with phosphates was found at the end of 107 days to be rooted almost twice as deeply as in similar soil to which no phosphates had been applied. Not only the quantity of nutrients but also the time at which they are absorbed affects quantity and quality of yield. In spring wheat and oats, the nitrogen or protein content has been shown to increase continuously as applications of nitrate fertilizer were made later and later in the life of the plant. Winter wheat and rye gave a similar response to applications made in the latest periods of growth. Thus, the amount of nutrients in the subsoil available to the younger and more vigorously absorbing roots is of great importance.

The problem of getting the phosphatic and potassic fertilizers, which do not leach extensively, into the deeper soil where they may be more efficient is one with which students of fertilizer practice should be concerned. Likewise, the time and quantity of the applications of nitrates, which leach freely into the deeper soil, in relation to rainfall and root depth is a field needing investigation. It is possible that methods of tillage and cultivation can be modified so as to give, in advance of the crop season, a supply of nitrates which might leach into the deeper subsoil in time for their absorption at the most effective period in the development of the crop. In the semiarid wheat growing regions of eastern Washington, the nitrates developed in the summer fallow of the preceding year are found in the subsoil the following spring at a depth of 4 feet or more.

Excellent results are reported from the use of subsurface application of fertilizers in growing potatoes in New York. The fertilizer is broadcast and thoroughly mixed with the upper 4 to 5 inches of soil by means of a heavy disk harrow. The field is then replowed 9 to 10 inches deep before planting. Severe drought may so thoroughly dry the upper layer of soil that it is of little use to the plant. However, the fertilizer well mixed in the deeper soil has, by this time, induced such a heavy root development that the crop continues growth unchecked throughout the drought period.

In other instances, especially in orchard growing, placing fertilizer deep in the subsoil has been shown to be an excellent practice. Until very recently, however, the deep-rooting habits of crop plants have not been appreciated by students of fertilizer practice nor have the nutrients of the subsoil greatly interested them. Hill fertilizing of corn promotes more vigorous early vegetative development and earlier tasselling and earing.

The observation of farmers that corn fertilized in the hill sometimes suffers more from drought than when grown in soil where the fertilizer has been uniformly distributed may be explained by a study of root extent in relation to tops. Although no differences were found in the actual abundance, depth, or lateral spread of the roots, the more luxuriant plants resulting from hill fertilizing had a relatively smaller root system. This may also explain why, in Missouri, applying fertilizer in the hill or row yields good returns during seasons of abundant rainfall, but in dry seasons, there is more danger that the fertilizer may cause the corn to "fire" than when it is applied ahead of the planter with a fertilizer drill.

Because of the extensive development of the roots of practically all cultivated plants, it seems probable that the chief effect of hill manuring is to promote vigorous early growth and that the plant receives little benefit from the manure at the time when it is completing its growth and maturing its seed. It has been shown by means of water cultures that when only a part of the root system is supplied with one mineral element and the entire root system with all of the other necessary elements, the plant does not absorb as much of the one element as it would were this element available to all of the roots.

The fewer the roots supplied with the element, the smaller the total amount absorbed, although the amount of the element absorbed per gram of root increases greatly as the number of roots in the complete solution is diminished. This applies when the total amount of the element is equal to or in excess of the needs of the plant.

For example, with nitrogen and phosphorus, the total amount absorbed by plants with half their roots in the complete solution was only 76 per cent of that absorbed by plants with all their roots in the complete solution.

Nor did increasing the element in question in the complete solution appreciably alter the results. Since a plant is unable to attain a maximum absorption by means of only a portion of its root system, it is obviously important to apply fertilizers in such a way that, as far as possible, all of the roots of the plant will be supplied with all of the fertilizing elements. Since the lateral diffusion of fertilizer salts in the soil is small, this can best be done by distributing the fertilizer uniformly over the whole area occupied by the roots.

Role of Roots in Cultural Practices

Root distribution and development is greatly modified by various cultural practices, but our knowledge is very incomplete and a great deal more experimental work should be done in this field.

Transplanting

Nurserymen transplant trees and shrubs two or three times in order to force root development near the stem and thus to insure the preservation of more young roots when the plants are lifted for shipment. Hence, they have a better chance for recovery when again set out. This explains why nursery-grown trees and shrubs usually survive transplanting so much better than those secured from places where the roots have made their natural growth. Likewise, market gardeners find that transplanting young plants of cabbage, tomatoes, etc., while growing in cold frames, is a great advantage in assisting them to endure the final removal to open ground.

In the transplanting of trees, both depth and spacing should be given careful attention. If the roots are placed either too deep or too shallow, the plant is at a decided disadvantage. Proper spacing in forest planting is necessary to obtain well-balanced and wind-firm root systems. Not infrequently, orchard and shade trees as well as trees planted for windbreaks are so closely spaced that insufficient room for proper root development results in a marked decrease or actual cessation of growth. Layers of compact soil often play an important part in shaping the root system.

Western yellow pine seedlings transplanted in clay loam soils with their roots against one side of a hole showed a marked tendency to grow a one-sided root system, the growth being away from the side of the hole towards the looser soil within. When transplanted by the usual "trencher" method, the roots "invariably develop only in the plane corresponding to the longitudinal axis of the trench." In plowing for cultivated crops on heavy soils, the depth should be varied from year to year so that a too firm "plow sole" will not develop at a certain level and tend to confine root development to the plowed layer.

The time of transplanting and of early spring cultivation should be considered in relation to root, development. In the red currant, for example, which is representative of many other plants, root growth started in advance of bud development and before the beginning of the growth of the stem. Root growth in apple trees began a little earlier than did shoot growth and continued for several weeks. This gradually passed into a second period extending over a part of June and July during which shoot growth was still active and root growth relatively inactive, a condition which was most marked about midsummer.

Later as shoot growth slackened, root activity greatly increased and continued until late autumn. In fact, much of the larger part of the root growth was made late in the summer and autumn. In climates which are not too cold,... fall-transplanted trees are more likely to give a good stand than corresponding spring-set trees, for during the winter months, new root formation is initiated and water can be absorbed in the spring as fast as the new shoots and leaves use it.

The spring-set trees, on the other hand, must wait until new roots are formed before they can take up moisture, and if soil conditions remain unfavourable for their root formation and atmospheric conditions stimulate vegetative growth of the top, the pushing shoots will wilt and die, and the tree will be lost. In the autumn, conditions are favourable for root growth for some time after good growing conditions for the top have passed; in the spring, they frequently become favourable for top growth before or simultaneously with suitable growing conditions for the roots.

Balance between Root and Shoot

The maintenance of a proper balance between root and shoot is of very great importance. If either is too limited or too great in extent, the other will not thrive. The root must be sufficiently widespread to absorb enough water and nutrients for the stem and leaves, which, in turn, must manufacture sufficient food for the maintenance of the root system. It is a well established fact that grasses develop a better root system when they are mowed once or twice a year than when they are closely and frequently grazed.

In fact, one of the most important factors in the various systems of range and pasture management is to permit the seedling grasses to become well rooted before the tops are removed by grazing. In transplanting crops of various kinds, many of the roots of the young plants are necessarily destroyed. Hence, the top must be pruned back or the plant protected from excessive water loss until the balance between the absorbing and transpiring systems is reestablished. Conversely, a top too small to manufacture sufficient food to feed an extensive root system creates an unbalanced condition in the plant which, if uncorrected, may retard its development and may even result in death.

Pruning back the vines of sweet potatoes is a practice followed by some growers who believe that reducing the growth of foliage stimulates root development. Actually, root yields are reduced. Experiments with 3-year-old almond trees have shown that the development of both the top and root systems was inversely proportional to the severity of the pruning of the tops. The spread of the roots was over a third greater where the pruning was light than where it was severe. Heavy pruning of the first crop of indigo, leaving a few leaves, is said to have resulted in far less damage to the roots and nodules and in a much more rapid development of the second crop than pursuing the common practice (in Bihar, India) of completely cutting back the first growth.

Wheat grown in culture solutions poor in nitrogen develops a large root system which, when nitrogen fertilizer is added, absorbs much more of this nutrient than is needed for normal growth. This results in abundant tillering. More than four times as many tillers may develop as among plants with smaller root systems. This indicates a causal relation between differences in root-absorbing capacity and tillering. Buckwheat, beans, and other plants adapted to shade, sunlight, and an intermediate condition, respectively, have been grown in shade and sunlight. In all cases, plants grown in sunlight had a larger root system than those grown in shade, a result attributed to the increased transpiration.

As regards the roots of seedlings, it has been shown that their development is greatly influenced by the nature of the reserve food supply in the seed. The more nitrogen a seed contains, the greater is its shoot growth as compared to root. An abundance of carbohydrate foods and a somewhat limited nitrogen supply promote rapid root development.

Tillage Practice

The value of various depths of plowing, listing, or subsoiling, and the preparation of the seed bed, as well as the time, depth, and manner of subsequent cultivation are usually judged entirely by the increase or the decrease in growth and the yield of the aboveground parts. Too little attention has been given to the effect of these practices upon root activities and development. A knowledge of root development under each of the various methods of tillage, systems of mulching, etc., with their resultant effects upon water content, nutrients, aeration, and other edaphic factors, will not only give a logical cause for the results obtained by these practices but will form a scientific basis for the application of other methods or combinations which may result in greater yields.

It should be kept clearly in mind, however, that tillage methods, as such, may not influence root habits directly. Differences in growth and distribution of roots in the soil are in response to differences in physical and chemical conditions of the substratum in which they live. Tillage methods are merely the means of bringing about these changes. Much farming practice, both ancient and modern, finds explanation in root physiology. For example, loosening the soil by plowing results in the storage of water and in better aeration. This not only makes conditions more favourable for seed germination but affords better conditions for the growth of roots and for such soil organisms as nitrate bacteria.

The latter produce greater amounts of nitrates which, in turn, affect root growth. Chemical corrosion is promoted; that is, nutrients are more rapidly liberated from the soil particles, run-off is lessened, temperature conditions are changed, etc. All these factors affect root habit. Moreover, the loosened soil makes root penetration easier. Roots tend to develop a shorter and more compact structure in dense than in loose soils. Subsoiling carries the air still deeper and at the same time raises more minerals to the surface soil layers. It modifies in many ways (mostly unfavourably as measured in crop yield) the physical, chemical, and biological factors of the soil. All these changes are ultimately reflected in root habit.

Intertillage

The depth of intertillage exerts a marked effect upon root habit and often upon yield. For the highest yields, cultivation should never be deep enough to injure the roots seriously. They should be allowed to occupy the richest portion of the soil, which is usually the furrow slice. The proper type of cultivation is deep enough to kill the weeds but shallow enough to reduce root injury to a minimum. A decrease in the yield of corn of 2 to 8 bushels per acre was brought about by deep cultivation in Illinois and a decrease of 13 bushels in Missouri. Where the roots were pruned to a depth of 4 inches at a distance of 6 inches from the hill, the yield was decreased 17 bushels per acre.

When soil adjacent to hedge rows of Osage orange is not cultivated, their roots gain possession of a much larger area than where thorough cultivation is practiced. One investigation showed that lateral root extent in the two cases was 44 and 29 feet, respectively. Under conditions of cultivation, the water content of the deeper soil is increased, which, in turn, lessens the necessity for wide root spread. The ability of perennial crops like alfalfa and clover to grow in close proximity to the trees while annual crops do so with difficulty is due in a large measure to the more deeply penetrating roots of the perennial crops. That mulching the soil and lack of tillage result in a marked growth of fibrous roots in the surface layers has been repeatedly demonstrated.

Roots of fruit trees are often very superficially placed under a straw mulch. For example, apple trees grown for 8 years under a heavy straw mulch and under cultivation, in Indiana, had very different root habits. Quite in contrast to the depths of roots under cultivation was the shallow root system of the trees under the straw mulch. Here, they came very close to the surface of the soil. In fact, roots half an inch in diameter or larger were found growing on the surface of the soil under the straw mulch, and very many fibrous roots were found on the surface soil and penetrating the decaying material.

The roots, while more abundant than in the cultivated plots, averaged much smaller in diameter, and 75 to 80 per cent of the entire root system was in the surface foot of soil. Contrary to popular belief, deep tillage as a means of overcoming drought or of increasing yields has little foundation in fact. Plowing does not increase the water-holding capacity of the soil nor the area in which the roots may develop or from which plants may obtain food... Yields cannot be increased nor the effects of drought mitigated by tillage below the depth of ordinary plowing.

Crop Rotations, Cover Crops, and Intercropping

Crop rotations on different types of soil and under different climatic conditions should be worked out with reference to root relations. That "knowledge of root systems is the basis of agriculture" should be given more than casual consideration. It may be found practicable, especially in semiarid regions, to grow short-rooted and densely rooted crops alternately with those of longer and more spreading root systems. European investigators have emphasized the fact that crop rotations should be made with reference to root change.

In the semiarid regions of Russia, which, in many ways, are similar to our Great Plains, definite rotations involving this principle have been worked out. For example, deeply rooted and densely rooted winter wheat or rye is followed by the more meager- and shallow-rooted potato crop, and this by barley or oats, which, in turn, is succeeded by the shallower and more poorly rooted flax. This is considered to be an effective means for contending against drought, since it averts the perennial drying up of the root-inhabited soil layer.

In humid regions, under intensive agricultural conditions, two differently rooted crops may be grown in the same field at the same time. Nurse crops of oats for clover are common; pumpkins or soy beans are often grown with corn; and the use of cow peas with corn is an old practice, especially over the southern part of the corn-growing region. Growing mixed cultures is a common practice in India, where they usually outyield pure ones. The selection, breeding, and adaptation of crops for and semiarid regions should logically centre about their efficiency as absorbers and conservers of water.

Plasticity of root systems as to depth, lateral spread, degree of branching, etc., goes far towards determining the ability of a crop to make sufficient growth and yield to warrant its cultivation in dry lands. Moreover, under these conditions, root competition is an important factor in determining the rate of seeding. It seems not improbable that some of our best-yielding crops may be able to outstrip others, largely because of their greater efficiency in securing a larger and more constant supply of water and nutrients.

Why certain artificial mixtures of grasses and other herbs may thrive in pastures and meadows, while others do less well, must depend to a large degree upon competition of root systems. This is the case in native grassland, where it is usual for 200 to 250 individuals or groups of individual plants to grow in a single square yard, due to lessened competition resulting from absorption at different soil levels and to different periods of maximum aboveground activity during the growing season.

Depth of rooting can be controlled to a considerable extent by the use of cover crops or by intercropping. If the surface soil is depleted of its moisture by absorption, roots of both crops penetrate deeper. The effect that one kind of plant may have upon the root habit of another by modifying soil conditions is well illustrated in the case of chaparral and Monterey pine at Carmel, Cal. Here, the trees growing in an open stand among the shrubs died when the latter were cleared away. However, a new growth of pine flourished on the same area.

The chaparral had shaded the soil and lessened evaporation from its surface, and the dense layers of rootlets and accumulated humus held the moisture in the surface soil. Consequently, the trees were shallow rooted and died of drought when the protecting cover was removed and the soil desiccated. Seedling trees in the changed habitat evidently rooted more deeply.

In fruit orchards of Oregon, it was found that the cultural treatment to which an orchard has been subjected has a strong influence upon the location of the major portion of the fibrous roots of the trees. Where clean culture had been practiced without the use of the plough, a thick mat of fibrous roots was found immediately below the soil mulch. Few roots extended to depths of 8 to 12 inches below the mulch. But in a few restricted areas that received neither cultivation nor irrigation, the roots were found to be distributed from near the surface to a depth of 12 to 16 inches.

Under sod and irrigation, the roots were quite uniformly distributed from near the surface to 2.5 feet in depth. Under the loose surface soil of the cultivated area, there had been formed an impervious hardpan which was entirely absent in the untilled and unirrigated land. It has been observed by several investigators that trees growing in competition with grass have a relatively heavier root system in proportion to tops as compared with those under cultivation. This relationship has been reversed, however, by adding nitrate fertilizer to the soil about the trees in the sod.

The presence of the roots of a previous crop in the soil where another crop is growing exerts an influence upon it in several ways. Upon their decay, they enrich the subsoil to a depth of several feet. A network of tortuous channels, formerly occupied by the roots and about which the soil has been more or less compacted by lateral pressure exerted in their growth, fills soil and subsoil. This permits better aeration which may be an important factor in the oxidation of harmful substances originating from the decay of the roots.

The ancient practice of letting a soil lie fallow for a time may restore its productivity, either by allowing time for the formation and diffusion of more soluble mineral salts, or through the removal by leaching or oxidation of injurious substances formed by roots. It has been clearly demonstrated that the presence of sorghum roots and stubble has a distinctly depressing effect upon the yield of wheat. Preliminary field and pot tests with tobacco indicate that the injurious effects of preceding crop plants come mostly from the roots rather than the tops of these plants.

Roots of potatoes, hairy vetch, and corn retarded the growth of tobacco even when their aboveground parts were removed from the field in harvesting. Rotation of crops may derive its value both from the different demands made by various crops upon the nutrient supply of the soil and from the fact that organic materials added by decaying roots are often less injurious to other kinds of plants than to the crops producing them. Indeed, it is held by some that the major benefit afforded by the addition of fertilizers is not the replacement of mineral salts removed by the crop but their beneficial effect, in neutralizing unfavourable soil conditions introduced by root excretion and decay.

Other Root Habits

There are many other ways in which root habits of plants are related to problems connected with crop production. Prominent among these are problems of soil erosion and weed eradication, as well as crop production on alkali or acid soils. An other important subject is the relation of root development to disease in plants.

Soil Erosion

Closely connected with crop rotations is the problem of soil erosion, whether by wind or water. This holding of the soil for cultivation applies equally to plants of pasture and meadow lands and to annual crops. Nearly half of the United States—the hilly half—is being seriously impaired by water erosion. Indeed, erosion is one of the most serious dangers that threaten the agricultural and pasture lands of the nation.

The roots of plants are efficient soil binders, the effect being more continuous with perennials. Among cultivated crops, this is especially true of meadow and pasture grasses. Even though they may not form a continuous sod, the soil particles are held in place by the extensive and minutely branched fibrous root systems. Grasses that spread by rhizomes and form a compact sod are especially efficient. Not only do the underground stems and roots hold the soil, but the tops form a complete surface cover which efficiently shields the soil from the destructive action of wind and water.

Removal of the cover of vegetation, whether by overgrazing or plowing, results in root decay and is frequently followed by soil erosion. Even after death and decay, roots and tops exert an important effect in cementing together the soil particles as well as by absorbing and holding several times their own weight of water. But once the soil is exposed, every drop of rain that falls has the power of removing soil particles and with them the soluble salts essential to plant growth. Some of the most important grasses of sandy areas are of little or no forage value but exert a pronounced effect upon stabilizing the loose soils and thus permitting the growth of other vegetation. Certain crops like rye are also efficient soil binders.

Weed Eradication

In keeping the soil free from noxious weeds, especially

perennials, a knowledge of the position, extent, and growth habits of the underground parts is very important. Here as among most groups of plants, however, few data are available. Why some species are more detrimental to crops than are others may be due in part to their absorbing water and nutrients from the same level as the cultivated plants.

Marked differences occur in different species. The roots of the dogbane (Apocynum androsaemifolium), for example, fill the soil rather completely to a depth of 3 to 4 feet. The vervain (Verbena stricta), common in old pastures, occupies the soil in a similar manner but the cord-like roots of the ironweed (Vernonia baldwinii), also frequent in grassland, reach depths of 9 to 11 feet and branch rarely or not at all until they enter the third or fourth foot of soil.

Relation to Alkali and Acid Soils

Studies of root systems in soils impregnated with alkali point out clearly that the adaptation of plant to habitat is often largely one of root distribution above or below the layers of greatest salt concentration. Shallow-rooted native species which cannot endure alkali may grow upon land which contains alkali at depths below those of root penetration. Orchards and vineyards are sometimes planted in soils of rather high salt content, and the root systems may become thoroughly established in a non-toxic lower layer of soil which is less alkaline than the surface layers.

Shallow-rooted crops may fail to give a satisfactory yield because the alkali tends to concentrate near the surface if evaporation is great. "This accumulation makes the salts very strong throughout the feeding zone of the plant and, therefore, toxic even when the total quantity of salts in the upper 3 or 4 feet is rather small." Winter wheat is a poor crop on alkali land, because it is seeded in the fall when the alkali is usually concentrated near the surface, although experiments show it to be more tolerant than many spring crops which, in practice, are better adapted to alkali lands than is wheat.

Deep-rooted plants, like alfalfa and trees, may penetrate the alkali strata by growing in the upper soil while the alkali is beneath and gradually feeding lower as the alkali accumulates at the surface. In this way, some plants not exceptionally tolerant may withstand what seem to be excessive quantities when the whole feeding zone is not considered. Where alfalfa, cotton, and other deep-rooted plants get a good start but encounter a strong alkali stratum at a short distance below, these plants may prove less resistant than the cereals which may feed in the upper less alkaline soil.

The sensitiveness of seedlings of sugar beets and alfalfa to alkali and the relative alkali tolerance of well-established plants have long been known. Heavy rains or irrigation during germination and establishment may so dilute the harmful salts that the seedling stage may be safely passed. Both crops soon develop a deep root system. The alfalfa plants shade the soil and thus hinder surface evaporation and salt concentration, Moreover absorption from the deeper soil layers also retards the upward movement of water and solutes.

The high resistance of alfalfa to alkali, once the plants are established, is due in a very large measure to its deep root system which often absorbs in the deeper soils below the salts. In the case of sugar beets and other intertilled crops, the surface mulch resulting from cultivation, aided by the shade produced by the crop, tends to check the rise of the salts. Beet roots absorb at all depths to 6 feet, and the plants thrive even when the alkali present in the surface soil is very great. It has been demonstrated that plants with extensive root systems are less subject to injury from harmful salts than those whose root systems are more poorly developed.

In problems of crop production on acid soils, if attacked from the more modern and logical viewpoint of lime requirement of the plant rather than that of the soil, the root system again plays a decisive rôle The lime requirement of the plant is determined not only by its lime content and rate of growth but especially by its feeding power for lime. The latter is in proportion to the character and extent of the root system, the internal acidity of roots, and their excretion of carbonic acid. Extensive investigations on the effect of soil acidity on root development have been made on moor soils in Europe.

The roots of cultivated plants were found to penetrate into the soil only as deeply as the addition of basic materials had sufficiently freed the soil of free acids. Certain native species were found to be less sensitive, their roots occurring in the deeper soil. Experiments with potted soils showed that the length of the root system agreed very closely with the depth of the acid-freed root bed. Studies of root development in the field, where the deeper soil layers were freed from acid by the addition of lime, confirmed the pot experiments.

Wheat seedlings grown in nutrient solutions with a high H-ion concentration developed root systems which were abnormal in being short, stubby, and much branched. The protoplasm of the root hairs was found to be coagulated and flocculated and the hairs were probably rendered ineffective as absorbing organs. Excessive acidity affects the roots of crops by partially or wholly retarding growth in length. The roots often thicken and soon become dull white in colour. Sometimes, as in the case of root rot of conifers, injury to the roots resulting from acidity of the soil leads to infection by fungi which cause root decay.

Aeration and Soil Temperature

Some fundamental relations have been recently worked out between rate of root growth, aeration, and soil temperature. Under normal conditions of aeration, the rate of root growth is known to be influenced by soil temperatures in such a manner that there are three welldefined temperatures for growth. These are the maximum or highest temperature at which root growth is possible; the optimum, at which temperature growth is most rapid; and the minimum, below which it ceases. But under a diminished oxygen supply, these cardinal temperatures seem to be greatly modified.

As the oxygen supply in the soil air is decreased, rate of growth diminishes in a soil with a high temperature. For example, corn roots, in a soil atmosphere of 96.4 per cent nitrogen and only 3.6 per cent oxygen, at a temperature of 30°C. grow about one-third as rapidly as at the same temperature under normal conditions of aeration. But at 18°C., growth is increased to about two-thirds the normal rate at that temperature when the soil is well aerated. Similar results hold for cotton and other species. Such crops, to attain a fair rate of growth at a time of high soil temperatures, must be in a well-aerated soil; otherwise, the rate of growth is considerably reduced.

The roots of cactus (Opuntia) are of two kinds, namely those which grow directly downward from the base of the shoot and apparently serve the purpose of anchoring the plant, and those which spread laterally in the shallower soil and appear to function mainly as absorbers of water and nutrients. An experiment was so arranged that cuttings of Opuntia versicolor grew for two years in an adobe soil, or in sand filling a large pit dug in the adobe, or on the edge of the pit in both adobe and sand. A suitable water content was maintained in both soils during the growing' season.

Owing to the differences in the soil texture the sand was better aerated at all times than the adobe. Temperature differences at corresponding depths probably also occurred. The plants which were in the adobe soil had a normal type of root development, the absorbing roots reaching out far beyond the shoot but lying near the surface of the soil. On plants grown in the sand the differentiation of the roots into the two systems was obscured.

There were no proper superficial roots and no proper anchoring roots but the roots extended downward at many angles and divided the soil mass encompassed by them fairly equally. But the roots of the plants situated on the line separating sand and adobe were of both types. Those in the adobe were either superficial and horizontal or extended rather directly downward, and those on the sand side developed after the manner of the roots wholly within the sand. This shows that the roots of Opuntia are exceedingly plastic and are directly affected by aeration and soil temperature.

Experiments with the development of nodules on legumes have shown that they become much larger when soil temperatures are most favourable. In the soy bean, a consistent increase in dry weight of nodules occurred as the soil temperature increased from 15° to 24°C. At higher temperatures, a progressive decrease occurred. Alfalfa, red clover, and field peas likewise gave a maximum nodule production at a soil temperature of about 24°C.

Plant Disease

The relation of root development to disease resistance of crops is important and requires intensive study. Recent investigations have demonstrated a close relation between the vigor of root growth and disease resistance. "Water stress" in cotton affects most seriously the plants with the greatest vegetative growth. They remain longer in a wilted condition between irrigations and show an earlier recurrence of wilting after irrigation. Little difference has been found in the size of the root system between large and small plants, and this seems to explain the behaviour of those with too large aboveground parts.

In some species, similar conditions promote shedding of leaves. For example, the pine needle shedding disease can be, cured by promoting the development of a good root system. Where the disease is physiological, resulting from malnutrition or insufficient aeration, this relation is at once apparent. In some cases, defective soil aeration causes disease. The wilt disease of Java indigo and other monsoon crops of India is due to damage of the root system resulting from defective aeration. This is brought about by the rise of the water table, combined with the decrease of the porosity of the surface soil by heavy rains.

Wilted plants possess few fine roots and nodules in an active condition, most of them being dead or discoloured. Deeply rooted varieties of crops are especially subject to wilt, while shallow-rooted ones are little affected. The root tips often show marked aerotropism and grow upwards towards the air. In some cases, they abandon the soil and grow over the surface of the ground. Sometimes, insects and fungi attack the crops and cause disease, the actual attack following the operation of some factor such as poor soil aeration or soil temperatures unfavourable for good growth and other functions of the root system. This is thought to be due to change in cell sap, arising from root damage, which prepares the way for the parasite.

Certain plant diseases that cause enormous economic losses are due to organisms that enter the root and cause it to decay entirely or in part, thus bringing about reduced yields or total loss of aboveground parts. In the root rot of clover, tobacco, and other crops, the root system is partially or entirely destroyed by various soil-inhabiting fungi. Sometimes, as in club root of cabbage due to the invasions of a slime mould, the parts belowground are greatly deformed. The root rot of tobacco is marked by the stunting of plants in various degrees due to a reduced root system.

The extent of the damage is determined in a large measure by the environmental conditions surrounding the roots of the host. Diseases such as flax wilt and cabbage yellows are caused by fungi entering the root hairs, pushing back through the cortical tissues, and growing throughout the vascular system. These invasions of root and stem result in diminution of water and supplies of food materials from the soil which, in turn, give rise to stunted plants. The host may be killed in the seedling stage or wilt and die at any time during its growth. In cabbage, the leaves have a pale, lifeless, yellow colour.

Sometimes, only one side of the root system is seriously attacked. Then the opposite side of the plant grows more rapidly and brings about a curving of stem and leaves. The invaded plants begin early to shed their lower leaves while making a weak attempt to continue growth above. Both root and stem are greatly dwarfed. The majority of the diseased plants continue a sickly existence for a month or more and then succumb. The Texas root rot of cotton, which also affects many other plants, is almost entirely confined to the roots.

The plants look normal and healthy, but wilting occurs suddenly. The entire foliage droops and dies and soon falls. An examination of the roots of freshly wilted cotton plants shows that the woody portions lying immediately beneath the cambium are deeply discoloured where the fungus has killed the tissue, probably by the secretion of a toxic substance of enzymic origin. The organism spreads underground by contact of infected roots of one plant with adjoining healthy ones of another. Thus, early planting and close planting of susceptible hosts greatly influence the amount of Texas root rot during a favourable season, because of better contact of roots underground.

A well-developed root system is a factor which greatly influences the summer spread of root rot. More recent investigations, however, where the root systems of cotton and alfalfa of both healthy and diseased plants were thoroughly examined, lead to a different conclusion. It was found that the fungus causing the disease does not spread from plant to plant by underground contact of diseased roots with healthy ones. It progresses through the soil, and the plants are attacked at any point on the taproot within the first foot of soil, and the laterals rot at the point of attachment with the taproot. These illustrations are sufficient to indicate the need of a thorough knowledge of the root habits of cultivated plants in combating plant disease produced by soil-borne parasites.

A more complete knowledge of the soil environment as it affects root growth and the growth of organisms attacking roots is of great scientific and economic importance. Increased attention to the relation of environment to disease inception and development is recognized as imperative to progress in plant disease control. Disease resistance and predisposition to disease may be largely dependent upon environmental conditions under which the plant is developing. Selecting plants whose roots are resistant to fungus attack or ensuing injury is a very promising method of procedure. Why they are immune is a problem that will require a thorough acquaintance with the structural changes and chemical composition of the roots.

2

Breeding and Seed Production

Root vegetables are underground plant parts used as vegetables. They are called root vegetables for lack of a better generic term, but include both true roots such as tuberous roots and taproots, as well as non-roots such as tubers, rhizomes, corms, and bulbs. Several types contain both taproot and hypocotyl tissue, and it may be difficult to distinguish the two.

Regardless of anatomical type, root vegetables are generally storage organs, enlarged to store energy in the form of carbohydrates. They differ in the concentration and the balance between sugars, starches, and other types of carbohydrate.

Selection of a seed crop that will thrive in the environmental conditions of the selected area is critical for achieving economic profit for the seed produced. Of particular economic importance are those with a high carbohydrate concentration in the form of starch. These starchy root vegetables are important staple foods, particularly in tropical regions. They overshadow the cereals throughout much of West Africa, Central Africa, and Oceania, where they are used directly or mashed to make foufou or poi.

Rice Breeding

Acetic Acid Treatment of Germination of Rice Seed

Reduction of germination

Rains and typhoons are prevalent during the rice harvest period in Taiwan. This condition causes germination on grain panicles and decreases rice yield and quality.

Dormant variety selection

To reduce the damages, dormant varieties were selected from breeding materials and then evaluated. These dormant seeds were not able to generate uniform rice seedlings when used for growing seedlings in the second cropping season

Generally, farmers keep rice seeds under room air temperature for 2-3 weeks to relieve seed dormancy. However, it becomes too late to transplant the seedlings with this method. Treating the dormant seeds with 50mM acetic acid for 24 h improved germination capacity by above 90% and produced uniform seedlings.

Anther Culture to Breed Aromatic Rice

The advent of various biotechnological techniques in recent years has offered new tools for developing high yielding rice varieties possessing good grain characteristics, such as aroma, to boost rice export and meet domestic demand. Among these tools, anther culture is emerging as an effective technique in rice breeding, particularly when applied to potential hybrid.

Arinatuc Rice Varieties Used as Parents

Four aromatic varieties were used as parents: DS 20, IR 68144, Kloong Luang 1, and Suphan Buri. Eight F_1

hybrids were selected for anther culture. N6 or MS were used as basic medium for anther culture. Panicles were collected prior to emergence from flag leaf. Cold treatment of panicles was carried out at 10°C for 7 days in the incubator. After cold treatment, the spikelets were surface sterilized with alcohol and $HgCl_2$. Anthers were taken out aseptically from the spikelets and placed on the callus induction medium. The cultures were incubated in the dark at 20°C. When the calli emerging from the anther lobe grew, they were transferred to the regeneration medium and placed under continuous fluorescent light for plant regeneration. A_0 plants were observed and collected to advance to A_1 generation and selected for A_2 generation.

Rice Breeding for Sustainable Agriculture

During the past three decades, plant breeders have greatly contributed to the development of high-yielding crop varieties to meet the food needs of a growing population. Breeders have changed the morphology and physiology of crop plants, and incorporated desirable traits and resistant gene(s) into traditional varieties while stabilizing or increasing crop production. In most rice-growing countries, the primary breeding objective has been a high yield potential.

The incorporation of the semidwarf gene into traditional tall, leafy rice varieties has made possible dramatic yield increases in the major rice growing countries (Table 1). The semidwarf rice variety IR 36 and other IR lines have been intensively used in national breeding programs for the development of high-yielding varieties. These are now planted in 60% of the world's rice land. In Korea, Tongil-type varieties derived from crosses between IR lines and Japonica varieties give yields 30% higher than the leading traditional varieties.

Table 1. Rough rice yield in selected Asian countries, World Rice Statistics

Country	Yield ton/ha			% increase over 1970	
	1970	1977	1987	in 1977	in 1987
Bangladesh	1.69	1.94	2.24	115	133
China	3.42	3.64	5.41	106	158
India	1.68	1.96	2.29	117	136
Indonesia	2.38	2.79	4.04	117	170
Japan	5.63	6.17	6.19	110	110
South Korea	4.55	6.79	6.02	149	132
Myanmar	1.70	1.95	3.04	115	179
Philippines	1.75	2.00	2.62	114	150
Thailand	2.02	1.59	2.02	79	100
Vietnam	2.15	1.94	2.70	90	126

Table 2. Comparison of grain yield and filled spikelet percentage, with staggered heading of three Tongil-type varieties/lines

Variety/line	Grain yield (mt/ha)[1]	Filled spikelets[1] (%)	Days from initial to full heading[2]
Iri 313	7.68	78	17 - 20
Tongil	7.13	71	13 - 16
Yushin	6.58	51	6 - 8

1) Average under two nitrogen levels

2) Estimated through the field observation

High-yielding varieties have made a great contribution to the world's food supply, but they also have several major problems. The high yields of these varieties can only be attained with a high level of inputs, in particular heavy applications of fertilizer (Fig. 1). This has led to problems associated with pest outbreaks in certain areas, while increased rice production has resulted in lower rice prices.

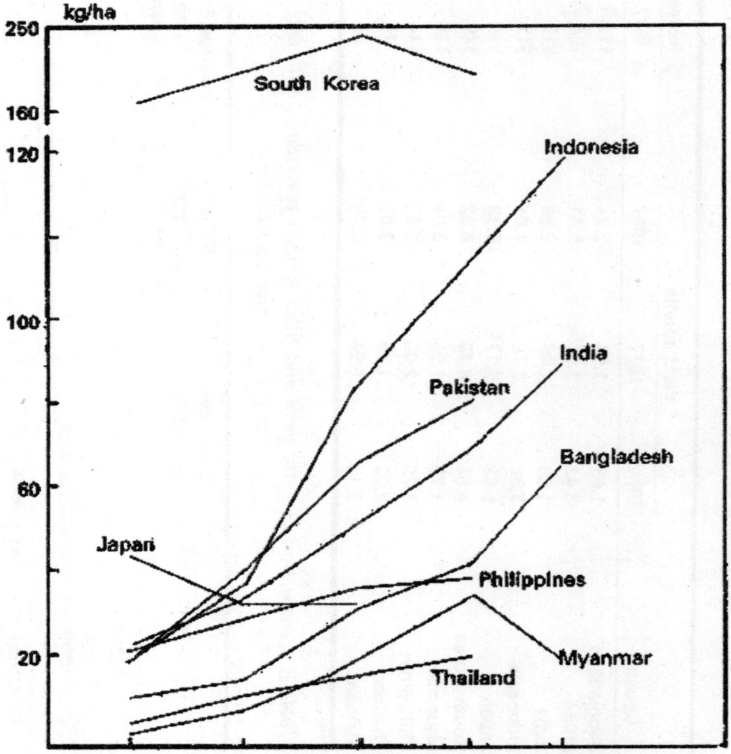

Figure 1. Consumption of fertilizer (N, P_2O_5, K_2O) applied to rice,

Although numerous improved high-yielding varieties have been released to farmers over the past few decades, little attention has been paid in rice varietal improvement to any concept of sustainable rice production. Particularly in countries like Korea, which are self-sufficient in rice, it is time for rice breeding strategies and objectives to change their focus and emphasize sustainable rice production which is appropriate to local farming systems and their economic and social background.

Breeding Strategies

The ultimate goal of crop breeding is to develop varieties with a high yield potential and desirable agronomic characteristics. In rice breeding, the most important qualities sought by breeders have been high yield potential; resistance to major diseases and insects; and improved grain and eating quality. However, there seems to be some conflict between these aims. Emphasis on high grain quality tends to result in unstable yields.

Conversely, too much emphasis on disease and insect resistance and stable yields leads to poor grain quality. Hence, breeding efforts should concentrate on varieties with the potential to minimize yield losses under unfavorable conditions, and to maximize yields when conditions are favorable. The following breeding approaches should be emphasized in producing varieties for sustainable rice production.

High Yield Production under Low Inputs

For sustainable agricultural production, attention must focus on the development of high-yielding crop varieties which have premium quality under low inputs. At present, breeders are examining a wide range of genetic resources towards this end.

Experiments have indicated that average yields of current rice varieties given no nitrogen application are only 70% of those given an optimum N application of 150 kg N/ha (Fig. 2).

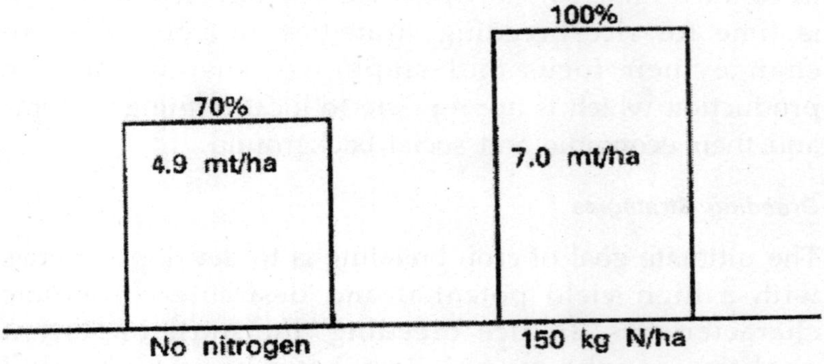

Figure 2. Average grain yield production under no nitrogen and 150 kg ha

However, F_1 hybrids are one possible way in which we may achieve yield increases with less applied nitrogen in the near future. A study of the growth of rice hybrids given five levels of nitrogen fertilizer showed that positive heterobetiosis in plant weight increased linearly from no nitrogen to 150 kg of nitrogen per hectare (Fig. 3).

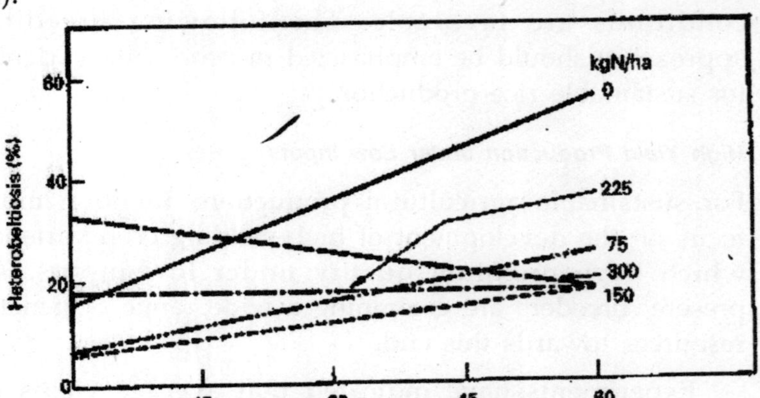

Figure 3. Change in heterobeltiosis on dry weight 60 days after.

The highest heterobeltiosis in plant weight was found when no nitrogen was applied, followed by 75 kg/ha and 150 kg/ha of nitrogen. In a comparative study of nitrogen response in F1 hybrids and conventional semidwarf varieties, F1 hybrids were observed to have higher yields when no nitrogen was applied (Fig. 4). This suggests that hybrids attain a higher yield at lower nitrogen applications than do conventional semidwarf varieties. In Mainland China, about 14 million ha are currently planted in hybrid rices, but the use of hybrids in other areas would depend on whether seed production was economical, and on whether grain quality was found acceptable by consumers.

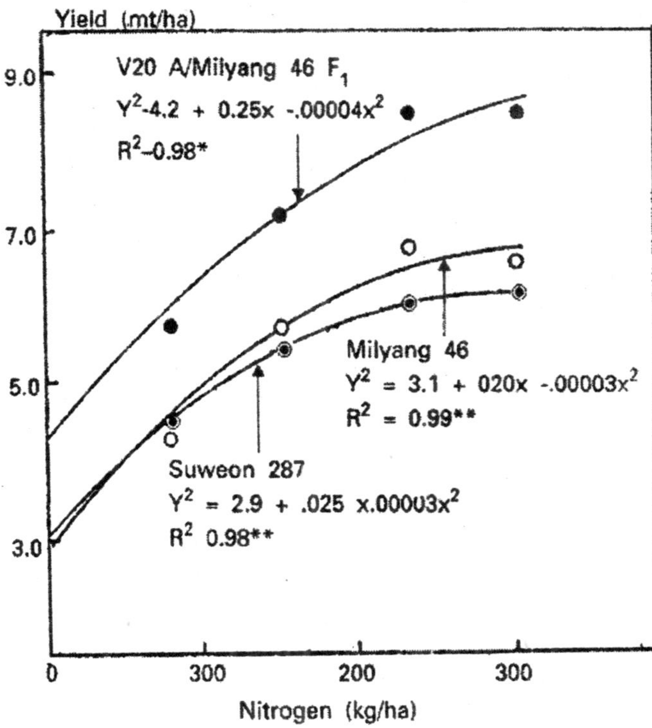

Figure 4. N responce of F1 rice hybrids and elite inbred cultivars for yield

Recently, photoperiod-sensitive (PGMS) and thermosensitive male sterility (TGMS) systems are giving promising results in seed production. It has been suggested that one approach to achieving high yields with low inputs may be to emphasize varieties with the characteristics of staggered heading during the flowering stage. A major barrier at present to even higher yields is that varieties with a large number of spikelets per panicle tend to have a lower percentage of filled spikelets, caused by nutritional competition between grains.

Staggered heading of tillers over a prolonged period, rather than simultaneous heading within a short period, may overcome nutritional competition between grains caused by an insufficient supply of carbohydrates as they simultaneously develop (Table 2). Furthermore, staggered flowering may avoid infection with disease and other unexpected stresses at the flowering stage. If heading covers a very long period, there may be management problems. Recently, the efforts of the International Rice Research Institute (IRRI) have been concentrated on modifying the plant type of modern high-yielding varieties in order to increase the yield potential. This new type has 4-5 productive tillers and large panicles, and is suitable for direct seeding.

There seems to be some difficulty in adapting current high-yielding varieties, which are very responsive to nitrogen, to a sustainable farming structure, but heterotic F1 hybrids and new plant types will be available in the near future and are likely to be much more suitable. Furthermore, high-yielding varieties can be expected to increase farmers' incomes, since they will permit more diversified cropping systems as less land is required for rice.

Insects and Disease Resistance

The host resistance to various biotic stresses is a very important aspect of high yields, and can be expected to play a significant role in sustainable rice production. Intensive efforts have been made over the years to incorporate resistant gene(s) into improved varieties, and there are now numerous varieties resistant to rice blast, bacterial blight, various virus diseases, and planthoppers. Some varieties possess multiple resistance to diseases and insects, but resistance in most varieties is controlled by a single gene and lasts only a few years, after which they become susceptible to serious disease or pest outbreaks.

For example, most Tongil-type varieties grown widely in Korea during the 1970s had a high level of resistance to rice blast disease, but in 1977 large areas planted in these varieties were severely damaged by blast. In contrast, varieties with the Xa 4 gene resistant to bacterial wilt have been grown in the Philippines for the last 15 years, and continue to be resistant. If a resistant variety becomes susceptible, a new variety with different gene(s) for resistance should be waiting ready for distribution to farmers. However, it is difficult to predict in advance which variety will suddenly need to be replaced, and it is difficult to release a new variety within a short period.

Furthermore, there may not be a wide enough range of genetic resources to provide a new variety with different genes. It is extremely difficult to identify polygenic resistance and incorporate it into improved germplasm. Current studies on host resistance to crops emphasize the durability of resistance . Some breeders believe that resistance to pests or other stresses is linked with low yield and poor grain quality. This relationship must be studied further.

Quality and High-yielding Varieties

Improvements in rice quality are very important in meeting the demands of consumers for healthy, high-quality food. Most rice breeders feel that improved high-yielding varieties in recent years have premium quality in terms of grain size, shape, appearance and palatability. However, consumers are still not satisfied with the grain quality of these varieties. There are many difficulties in evaluating consumer preferences, which vary widely from country to country and from person to person. However, consumers generally feel that even the best quality grain of improved high-yielding varieties is not as good as the best quality traditional varieties.

Many traditional varieties in both the tropics and the temperate zone have excellent cooking and eating quality, but a low grain yield. For many years, breeders have focused their attention on quality improvement, but without success. There seems to be some unknown genetic barrier to incorporating this trait into high-yielding varieties. At present, it is not possible to be very optimistic that we shall be able to develop high-yielding varieties with quality as good as that of traditional varieties, but our efforts continue, as does our work to develop high-yielding varieties with various grain characteristics suitable for processing.

Bambarra Groundnuts

Bambarra groundnuts are a traditional crop in Southeast Asia. They originated in the area around the edges of the Sahara desert in Africa, but have been grown in Southeast Asia for a long time. Countries in which they are found include Thailand, the Philippines, Malaysia and Indonesia.

In Africa, Bambarra groundnuts are a major leguminous food crop. As well as the seeds, the immature pods can be eaten as a kind of vegetable. In Asia, it has so far been grown only as a minor crop.

The Bambarra groundnut is a hardy crop, well adapted to high temperatures and dry conditions. In fact, it is one of the most drought-resistant of all the groundnuts.

Okinawa

Recently, scientists in Japan have been developing it as a potentiai crop for the subtropical islands in the south of Japan. These islands are known as the Ryukyu or Nansei Islands, the largest of which is Okinawa.

These islands experience a shortage of vegetables during the hot, dry summer months. Farmers are also looking for some new and profitable upland crops. It seems likely that the Bambarra groundnut may meet both requirements.

Three elite varieties were breed from germplasm introduced from Africa. Cultural practices were worked out: the best time to sow, fertilizer requirements and so on. Mulching was found to be effective in keeping down weeds and conserving the soil moisture.

Groundnut Cultivation

As for harvesting, it was found that the Bambarra groundnuts could be harvested by a small mechanized groundnut harvester already used by Japanese farmers. Although the Bambarra groundnut is sensitive to salty winds, covering the plants with a windbreak net was effective in alleviating the salt damage associated with typhoons.

Root wilt and virus could be controlled by breeding. Two-striped leaf beetle also attacked the plants, but was easily controlled by carbosulfan.

Consumers in Japan found that they liked the Bambarra groundnut very much. They especially liked a mixture of immature groundnut cooked together with sweet bean. We can expect more improvements in the future, as the Bambarra groundnut is adapted to the conditions of the southern Japan islands. Of particular importance is improved pest and disease control. In this way, an ancient minor crop may be changed to a new major crop for the islands.

Corn Seed Production

Corn is important in Asia as a livestock feed and a staple food, as well as a raw material for starches and sugars used in food processing and other industries. The governments of many Asian countries have been making an effort to supply their farmers with cheap, high-quality seeds. However, the overwhelming superiority of hybrids developed in Western countries means that these continue to dominate corn seed markets in Asia. This makes the price of hybrid seeds rather expensive, while location-specific breeding objectives tend to be neglected.

Corn Grain for Feed and Food

Corn is also an important human food in the Philippines. About one-fifth of the population depends on 'white corn' as a staple grain. Over 60% of the seed planted for grain corn in the Philippines is supplied by traditional varieties. This is because costly hybrid seed does not ensure a higher yield or a higher profit on most farms in the Philippines, where corn production tends to be high-risk. Crops may suffer from water deficit and other stresses, typhoons are common, and many farmers cannot afford enough pesticides, mineral fertilizers and other inputs.

In spite of these constraints, the use of hybrid seeds has gradually been increasing in the Philippines over the past ten years, reaching 14% in 1999. Most of these hybrid seeds are supplied by private commercial companies. The companies develop and provide hybrids in close cooperation with national research institutes. Thus, corn seed research in the Philippines has been carried out in complementary mode by the Government and the private sector. Government institutes are concentrating on open-pollinated varieties. Genetic materials developed by the public sector, including the parental inbreds, are shared with the private sector. In Japan, Korea and Taiwan, government research institutes involved in breeding corn seem to have changed their focus from grain corn to forage corn.

Technology for Producing Healthy Seeds

Pathogens capable of causing systemic infections on their host plants can usually also be transmitted through vegetative propagation from infected mother plants to the young plants. If all the seedlings or planting materials are infected by this type of pathogen, there may be severe losses of the crop after planting. Infected seedlings or planting materials can easily be transported to new areas, resulting in the invasion and establishment of new diseases. In areas where the pathogens already exist, infected seedlings or planting materials serve as inoculum sources of pathogens that can be transmitted by insects and other vectors, causing outbreaks of disease.

Because these pathogens are always located inside the cells and systemically spread throughout the plant, the diseases they cause are often very difficult to control by conventional measures such as pesticide sprays. Experience over several decades tells us that the most effective way of dealing with this problem is to produce pathogen-free seedlings or planting materials.

In Taiwan, the first propagation system for virus-free planting materials was established in the early 1970s, for the mass production of virus-free seed potatoes. It has been in operation for more than 26 years, and has proved to be effective in controlling potato virus diseases. Since then, more than nine more crops have been included in pathogen-free propagation programs.

Before the 1980s, the most widely used pathogen detection techniques were antibody-based, such as enzyme-linked immuno-sorbent assay (ELISA). These techniques are still widely used in laboratories, and have also been adapted to be used on an industrial scale. The reason for their wide acceptance by industry lies in the fact that their sensitivity and specificity are at an acceptable and reasonable level. Furthermore, the results can be reproduced, while there is a minimum of false positive and negative results. In all, ELISA testing is efficient and cost effective.

However, the invention of DNA-based technologies in the 1980s has driven most research on pathogen detection to take a new direction. Numerous reports have shown that there are many more DNA-based pathogen detection techniques than serological tests such as ELISA. Nevertheless, most of the DNA techniques are so far still confined to laboratory use. More research is needed for them to become usable on a routine, daily basis, for large-scale applications.

Types of Transmissible Pathogens

Pathogens which can induce systemic infection in their

host plants include viruses, viroids, phytoplasmas and some bacteria. As mentioned above, these pathogens can be transmitted from infected mother plants to seedlings, or to planting materials such as scions, cuttings, bulbs, and tubers through vegetative propagation. In fact, vegetative propagation by tissue culture, which has become the dominant method of propagating numerous plant species, is considered to be the most efficient way of multiplying and disseminating planting materials infected with systemic pathogens.

Viruses

Viruses are submicroscopic pathogens that cause disease in all known organisms, from bacteria to human beings. They are obligate parasites, requiring the presence of living cells for their multiplication. Recently, viral diseases such as influenza, polio, and AIDS in human beings, and foot-and-mouth disease in cattle and other ruminants, has caused huge economic losses and had a major social impact. There have also been numerous records of plant virus diseases which have caused major economic losses. Plant viruses differ from other plant pathogens, not only in their size and morphology but also in their mode of infection and the way they multiply in host cells and are transmitted to other host plants.

Most plant viruses are very simple, possessing either RNA or DNA as the genome. This is encapsidated in a protein coat. The coat protein and genetic nucleic acid of a virus are synthesized separately in different sites in susceptible cells. At an appropriate time, they are assembled together to form progeny virus particles.

Plant viruses differ from other plant pathogens in being unable to liberate themselves from the infected host plant into the environment. Without the aid of vectors such as insects, mites, nematodes or fungi, they are also incapable of introducing themselves into their host plants through the epidermis or natural openings. However, once viruses have successfully infected the host plant, they can easily be transmitted in vegetatively propagated materials. Some plant viruses may also be transmitted from plant to plant by mechanical contact, while others may spread in infected seeds or pollen. Another characteristic of plant virus infection is that virus particles can spread through the plant, and can be detected in tissues or organs that are some distance away from the infection site. This ability to spread throughout the organism is known as systemic infection.

Although some viruses may induce distinct symptoms, most viruses show visible symptoms which are very similar to those caused by other viruses. Therefore, we cannot depend on visual inspection to identify the causal virus. Since plant viruses do not produce any structure outside the infected tissue, and are not liberated from the infected cell, different methods have been developed to detect their presence in infected cells.

Studies of the molecular biological characteristics of viruses have helped researchers to understand the function of genes in the viral genomes, and the strategies adopted by viruses for gene replication and expression of their gene products. Over the past two decades, a tremendous amount of information has been gathered on the sequence and function of viral genes. This information is now widely used for the quick identification of viruses by DNA-based technology.

Viroids

Viroids are the simplest and the most primitive of all the

plant pathogens. Numerous plant diseases are induced by viroids, many of them common, such as potato spindle tuber disease and citrus exocortis. Viroids differ from viruses in being composed of only a naked single strand of ribonucleic acid (ssRNA), without the protection of a protein coat. The RNA of viroids is made up of small, molecules either rod like or a closed circle, which are highly base-paired.

Potato spindle tuber viroids have been studied in detail. They possess a serial arrangement of 26 double-stranded segments, interrupted by bulging loops of varying sizes. This arrangement of the RNA molecule gives them a characteristic stability and flexibility. In general, viroids are highly stable against thermal denaturation treatment. They are readily transmissible by mechanical contact or injury, but not by any vectors.

Like viruses, viroids can induce systemic infection in their host plants. Symptoms induced by different viroids can usually be distinguishable, and can thus be useful for diagnosis. However, it is better to use modern techniques for detection and identification, instead of depending on visual inspection alone. Since viroids contain only RNA without a coat protein, serological methods cannot be used to detect or identify them. However, they can be detected by DNA-based technology such as electrophoresis, DNA hybridization or polymerase chain reaction (PCR).

Phytoplasmas and Spiroplasmas

Phytoplasmas used to be called mycoplasma-like organisms (MLOs). They are a special group of plant pathogens, similar to the mycoplasmas known to cause human and animal diseases. Mycoplasmas differ from bacteria in that they lack a cell wall or penicillin-binding sites. They are, therefore, resistant to penicillin, to which bacteria are sensitive. However, mycoplasmas are sensitive to tetracycline antibiotics.

Mycoplasmas have a triple-layered plasma membrane. Because they lack a cell wall, mycoplasma cells have a characteristic lack of rigidity, and appear as pleomophic bodies. Phytoplasmas, or MLOs, differ from animal mycoplasmas in that they cannot be isolated as pure cultures and grown in an artificial medium. However, because they are similar in morphology, and also in their sensitivity to whip-like tetracycline, these plant-infecting mycoplasmas were called MLOs until the 1990s, when 'phytoplasma' was accepted as the official name of this group of pathogens.

There is also a similar, related group of pathogens, the spiroplasmas. These also used to be considered MLOs until about 1970, when it was found that they could be cultured. They were also found to have a spiral movement in artificial medium. Like phytoplasmas and mycoplasmas, spiroplasmas are pleomorphic, lack a cell wall, and are sensitive to tetracycline.

Plant-infecting phytoplasmas and spiroplasmas are both found in the phloem tissue. Although they are known to induce systemic symptoms, they are usually present in a low concentration and are not distributed evenly throughout the plant. Syndromes induced by phytoplasmas and spiroplasmas include yellowing, stunting, phyllody (development of leaves in places where ovules should develop), changes in color to reddish or purple and a bunching of shoots known as witch's broom. Although some viruses may also induce symptoms of yellowing and stunting, the latter three syndromes are generally typical of infection by phytoplasmas. Neither phytoplasmas not spidoplasmas are transmissible by mechanical contact. They can only be transmitted by vectors, particularly leafhoppers. A few can be transmitted by psyllids.

In the past, the identification of phytoplasma and spiroplasma diseases relied on microscopic observation, and biological techniques such as grafting or use of insects to inoculate susceptible hosts. In the 1980s and 1990s, serological techniques, especially monoclonal antibody technology, were applied successfully. More recently, molecular techniques including DNA probe hybridization and PCR, have been able to provide a quicker diagnosis.

Bacteria

Bacteria are prokaryotic microorganisms (i.e. they are cells that lack a membrane). Most of them are strictly saprophytic (i.e. they feed on dead organisms). They often help to decompose organic wastes in the environment. Around 1600 bacterial species are known to be pathogenic to plants. Apart from two species of Streptomyces, all plant-pathogenic bacteria are single rod-shaped cells. Most of the plant- pathogenic bacteria are motile (can move by themselves). They use flagellae to propel their bodies. These flagellae may be present singly, in groups, or distributed evenly over the cell surface. With the help of water, bacteria can move into host tissue through sites of injuries and through natural openings such as stomata. Most plant pathogenic bacteria induce local rather than systemic infection.

However, there are some specially evolved bacteria that can infect xylem or phloem tissue and become distributed throughout the plant, in a rather similar way to the systemic infection caused by viruses. In the early days, it was very difficult to culture plant pathogenic bacteria in artificial medium. For this reason, they were wrongly identified as mycoplasmas or MLO diseases for quite a long time. In the 1990s, it was found that most of them could be cultured in specially designed medium. They are now officially known as fastidious xylem- or phloem-limited bacteria, because they are found only in the xylem and phloem.

Bacteria belonging to these two groups give rise to diverse but usually species-specific symptoms in their host plants. Symptom inspection in the field is widely used for diagnosis. However, because of the low concentration of these bacteria in host tissue and the difficulty of culturing them, it is better to use serological and DNA-based techniques for quicker detection and identification.

Pathogen-free Planting Materials

The most effective way of controlling diseases that can be transmitted from mother plants to vegetatively propagated seedlings or planting materials is to develop a propagation system which ensures that all, or nearly all, of the plants produced are free of pathogens.

Developing a propagation system

The first step in developing a propagation system for disease-free materials is to get a complete picture of the pathogens that may attack the crop. To accomplish this needs several years of disease surveys, or consultation with experts. The key pathogens i.e. those which are most widely distributed or those which are responsible for major reductions in yield and/or quality, should be identified and targeted. Their biological properties, disease ecology, modes of transmission, methods of field diagnosis and keys for identification should be fully understood in as much detail as possible. This step is crucial for developing a sound structure for a propagation program that is effective in producing plants free from infection by key pathogens.

Some crops may be infected by numerous pathogens. In such cases, it may be very difficult to cover all of these, because of the cost. In real life situations, choices should be make to select key pathogens as the target to be eliminated from the propagation system.

Techniques for disease diagnosis and pathogen detection

The second step in developing a propagation program for healthy planting materials is to develop techniques for detecting the key pathogens. These techniques should be accurate, sensitive, and efficient enough to allow detection as early as possible, so the pathogens can be eliminated from the propagation system. Numerous techniques or diagnostic kits are described in the literature for the detection of plant pathogens and some are even available or the commercial market.

There are four main types of detection techniques.

- Biological inoculation;
- Microscopic observation;
- Serological detection; and
- DNA-based technology.

Of these four, biological inoculation is the most traditional, but is still a very accurate way of detecting pathogens. In this method, a susceptible plant host is deliberately infected with the pathogen. Biological inoculation requires a good system of different hosts which can be inoculated and observed for specific symptom expression. However, this technique requires space in a greenhouse and skilled labor to maintain the different hosts. It may take several days, or even several weeks, for symptom expression, which is usually too slow for commercial producers.

Nevertheless, biological inoculation reveals the live virus residing in the examined tissue. In contrast, the other three techniques may detect inactive viruses. This may give misleading results for certification. Therefore, biological inoculation is still used, in combination with the other techniques, in the certification of nuclear or mother stock plants. In this situation, there are usually relatively few plants to examine.

Like biological inoculation, microscopic observation requires experienced, well-trained staff and sophisticated facilities for processing. Again, it is best adopted to uses where the sample size is small rather than to large-scaled and routine inspections. Microscopic observation is therefore used mainly government institutes or specific diagnostic companies, rather than seedling producers.

It is the serological and DNA-based techniques, especially the former, which are currently widely accepted by the industry as the most efficient, dependable, reproducible and cost effective method of pathogen detection. There have been many improvements over the past two decades in serological techniques. For example, enzyme-linked immuno-sorbent assay (ELISA) has been modified into a semi-automated system that can process thousands of samples in one day.

There have also been tremendous improvements in DNA-based techniques in recent years. However, these are still used mainly by specialized laboratories with highly trained experts. Although DNA-based techniques are generally accepted as being more sensitive and specific for pathogen detection than the serological ones, they have some defects. These include the higher cost, the difficulty of reproducing results, and the possibility of false positives and cross-contamination between samples. These problems need to be solved before DNA-based techniques can be used as the routine basis of diagnosis.

Selection and maintenance of pathogen-free nuclear stock

The third step in the development of healthy plant propagation program is to select plants which are pathogen-free, and true to type with regard to the horticultural characteristics of the species or cultivar. Once selected, these plants should be protected from reinfection by pathogens. This is the time to decide which detection technique to use as the preliminary test to screen individual plants to make sure that they are free from infection by key pathogens. Because the number of plants for testing at this stage is usually large, techniques suitable for large-scale testing with acceptable sensitivity and cost effectiveness, such as ELISA, are desirable.

Once of this qualifying screening is completed, those selected individuals free from key pathogens should be collected and maintained in a vector-proof house for true to type selection and advanced pathogen testing. At this stage, since the number of plants for inspection should be reasonably small, more sensitive or accurate pathogen detection methods, even if they are more expensive in time and labor, can be used.

To be certain that the selected individual plants are free of all suspected pathogens, a combination of several techniques, including biological inoculation, microscopic observation, and serological and DNA-based methods are usually applied, either at the same time or in sequence. The cost, including the labor cost, is not usually the key concern at this stage of selection. It is accuracy which is the crucial point to be considered.

Furthermore, the selected mother stock plants should be carefully protected in order to prevent pathogens from entering their environment. Pathogen detection procedures should be conducted periodically to ensure that they remain free of pathogens. Every stock plant should be individually labeled, and kept well spaced from other plants to prevent mechanical contact. If there are insect vectors involved in the dissemination of pathogens, periodic pesticide applications should be made part of the management system. If re-infection is ever suspected, the suspect plant should be removed immediately from the area.

If no pathogen-free plants are available after extensive screening, this would indicate that the incidence of the pathogens concerned is already very high in the field. In this case, technology to eliminate pathogens from the mother stock plants should be used. Meristem tissue culture, thermo-inactivation treatment and many other treatments can be used for this purpose.

Pathogen-free Mother Plants

After you have established a dependable system for the maintenance of pathogen-free nuclear stock or mother plants, the next step is to consider how to mass propagate these stock plants into seedlings or planting materials for growers. One crucial point to be considered is the amplification factor. The reproduction rate of planting materials must be greater than the infection rate with the disease in the field. If the reproduction rate of pathogen-free planting materials is lower than the spread of disease in the field, the control effect of this strategy will not be acceptable to growers. Even more important, the reproduction method should be efficient, cost effective and acceptable to the industry. It cannot be considered only from the scientific point of view.

In the past, there were frequent cases of propagation methods being developed by researchers which were not accepted by commercial people, because of poor cost effectiveness and profitability. Tissue culture, for example, is a highly efficient method of reproducing planting materials. However, the cost, the high mutation rate and the possibility of re-contamination with pathogens may also be high. Where this is the case, a tissue culture system is unlikely to be acceptable to the industry or to growers.

Integration of Certification Programme

The experience of the Netherlands over several decades has shown that success in the use of healthy planting materials to control virus diseases relies at least partly on a successful certification system. This will help to give growers and importers confidence in the quality and pathogen-free status of the planting materials, so that they are willing to grow them. The fact is, the more pathogen-free planting materials are grown, the fewer the diseased plants which will exist in the fields, as they will be replaced by healthy ones. Consequently, outbreaks of disease will be suppressed.

If the government or any university could be responsible for inspecting the planting material propagation process, to guarantee the health of the plants and issue certificates, growers will probably be more than happy to grow these certified products, and abandon those may be infected. Most of those who use certified planting materials should be rewarded with better crops and less damage from disease. In fact, this system benefits not only growers but also propagators, who are more competitive if they produce certified planting materials.

However, most developed countries find that it is not easy to operate a certification system, not to mention developing ones. It takes good financial resources, trained manpower, good academic research and political support to carry out a certification system properly. Nevertheless, the experience of the Netherlands and of many other countries indicates that integrating the certification process with the propagation of pathogen-free planting materials will be the best guarantee of a successful disease control strategy.

3

Breeding for Growth and Yield

Selection of the more valuable species and geographic populations of crop plants and domestic animals has already been made, and the breeding practice mainly involves improving well-known varieties. By contrast, exploitation and reproduction of commercial forests mainly involve wild species, with only recent and minor use of strains selected for high yield.

Genetic selection to improve the yield of forest trees can be made at the levels of species, variety, provenance and individual tree. Such selection may be effective because native and naturalised species, varieties or provenances do not always fully exploit the managed environment of a region. They are usually adapted to survive the extremes of local climatic and edaphic conditions rather than to produce the highest yields of products required by man. Furthermore, they may not produce the desired type of wood. Where species are unsatisfactory for these or other reasons, yield can often be improved by proper introduction, testing and selection of the best adapted and most productive populations of exotic trees.

Selection of individuals for high yield can exploit the variation among trees within populations. It is most appropriate when the best provenance has been determined and when a large population from it is available locally for selection. Interspecific and interracial hybridisation have been used to improve the yield of forest trees and are promising methods of breeding in some circumstances.

Components of Yield

The total production of dry matter in a plant system is termed biological yield but only a portion, its economic yield is used by man. The dynamic relationship between biological yield (Y_{biol}) and economic yield (Y_{econ}) of a crop in a specific environment can be expressed as $Y_{biol} \times HI = Y_{econ}$ where HI is the harvest index. Breeding for increased Y_{econ} is possible if heritable variation occurs in Y_{biol}, or HI, or in both factors. The possibility of significant interaction of genetic and environmental components of Y_{econ} should be recognised.

Biological yield of a commercial forest depends on the inherent growth capacity of the genetic material being managed and on the environment in which it is cultivated. Examples of crop characteristics which may affect total organic production are leaf area and exposure, photosynthetic efficiency, relative growth rate, growth rhythm, rooting habit, ultimate height potential, health and response to plant density.

Harvest index, in a literal sense, is the ratio of the merchantable volume or weight of wood that can be taken from the forest to that produced therein. But this broad concept ignores the fact that only a portion of the wood harvested is recovered in the form of usable products.

The net harvest index achieved when a forest is exploited depends largely on the use of available wood, and on the efficiency of harvesting and processing. Considerable improvement of economic yield could often be obtained by greater attention to these aspects. But net harvest index is also influenced by morphological characteristics of the trees, through their effect on type of utilisation, ease of harvesting, and recovery of products.

Variable net harvest indices would be expected among populations and individual trees which differed in stem straightness, branch angle, stem taper, wood quality and distribution of photosynthate among stem, branch, bark, leaf, cone and root portions. For example, the finely branched and straight variety of *Pinus caribaea Mor.* from Cuba and of *Pinus pinaster* Aiton from Corsica must have higher net harvest indices than other varieties and races of the respective species. Pinus species/site interaction at Beerwah, Queensland, Australia, as shown by the superior economic yield of *Pinus caribaea Mor.* compared with *P. taeda L.* on red earth residual soils and by the much poorer yield of *P. caribaea* relative to *P. taeda* on humic soils.

Table 1. Biological Components of Net Economic Yield

Yield type	*Single trees*		*Stands*	
Yield component	*Volume*	*Weight*	*Volume*	*Weight*
Merchantable height	Yes	Yes	Yes	Yes
Basal area inside bark	Yes	Yes	Yes	Yes
Form factor	Yes	Yes	Yes	Yes
Stem quality factor	Yes	Yes	Yes	Yes
Wood quality factor	Yes	Yes	Yes	Yes
Basic density	No	Yes	No	Yes
Stand density factor.	No	No	Yes	Yes

The product $Y_{biol} \times HI_{net}$ may be termed net economic yield or recoverable yield. Net economic yield of a commercial forest is defined as the volume, weight or value of products obtainable from the wood produced per unit area of the forest during a rotation. Biological components of net economic yield are listed in Table 1. They are the broad criteria which could be used in direct selection for high yield.

Yield Selection

Importance of Species Selection

In many countries production by certain exotic species greatly surpasses that of most indigenous forest trees. For example, Wood has pointed out that in the United Kingdom introduced *Pseudotsuga taxifolia* (Poir.) Britt. yields in 50 years about twice the volume produced by plantations of the native *Quercus robur* L. in 100 years on sites of the same quality. There are numerous other examples of the advantageous use of exotic species in Europe, South America, Africa, Australia and New Zealand.

Often several native and introduced species can be established successfully in a forest region, but there may be great differences among them in biological yield, net harvest index and demand for their produce. This occurs most frequently when there are considerable variations of environment within the region, resulting in important interactions of species and sites. For example, in the coastal lowlands of southeastern Queensland, Australia, plantations of Pinus species are being established. Several soil types can be recognised on which three satisfactory species have given different responses after 15 years (Table 2).

Additional examples of species-site interactions are discussed by Jackson. He points out that proper delineation of major sites and correct allocation of species thereto are an effective means of increasing yield. In fact, if a species is not well adapted to the planting site, there is danger of economic losses sooner or later due to gradual or sudden environmental stress.

Table 2. Ranking of mean annual: volume increments of pinus species grown for 15 years on three soil types in southeastern queennsland

Species	*Red earth residuals*	*Lateritic podsolics*	*Humic soils*
P. caribaea Mor	1	1	3
P. elliottii Engelm	2	2	2
P. taeda L	3	2	1
Difference between ranks 1 and 2 (percent)	32	38	40

This ideal concept of selecting the best-adapted and most valuable species for each major site may raise problems such as choice of species, site delineation, establishment, the management and harvesting of discontinuous blocks of different species, and the marketing of several types of produce. On the other hand, use of several species in a region may be a valuable safeguard against pests and diseases and changes of product requirements in the future.

Provenance Selection

Genecological studies of many plant species have shown that characteristic local populations develop as a result of natural selection and other phenomena. Abundant evidence proves that genetically different geographic populations of most wide-ranging tree species vary greatly in economic yield at individual test sites.

As an illustration of the joint effect of provenance and planting locality on performance, the rank of several populations of *P. pinaster* for mean annual volume increment at a number of test localities in South Africa and Western Australia are shown in Table 3. The Portuguese race is favoured for planting because of its consistent high yield, but the Corsican race, characterised by superior stem form and branching habit (higher harvest index), may be valuable for interracial hybridisation.

Table 3. Ranking of mean annual volume increments of pinus pinaster aiton provenances grown at four localities in South Africa and four in Western Australia

Provenance	South Africa				Western Australia			
	A	B	C	D	A	B	C	D
Portuguese	1	1	1	1	1	1	1	1
Landes.	2	2	2	2	2	2	2	2
Esterel.	3	3	3	3	4	4	4	3
Corsican.	3	3	3	4	3	3	3	4

Actual gains in volume production obtainable by provenance selection have been estimated in a 14 - year old test of 14 provenances of *Pinus taeda* L. at Beerwah, Queensland, Australia. There was a difference of 38 percent between best and second best populations. The best population produced twice as much volume as the worst.

Problems in Provenance Selection

The risk of using exotic species or provenances when it is uncertain whether they are adapted to local environments:

1. They may succumb to unusual extremes of the environment.
2. Their growth rate and health may decline markedly after a good beginning due to unsuspected inadequacies of the site; the decline may not be apparent until a second rotation.
3. They may be attacked by insects or pathogens, especially if grown on marginal sites.
4. The type or quality of wood produced may be disappointing.

A dynamic, multistage programme of introduction and testing is imperative if local species are unsatisfactory. It is often impractical to test exotic species and provenances for a whole rotation before beginning establishment of commercial plantations. Many species and several provenances of each must therefore be established initially; most of them will be rejected in early or late stages of testing. A few will be advanced gradually to the status of major commercial species.

Problems involved in introducing and testing species and provenances are initial choice, early tests, field trials and effective cooperation. The initial choice of species and provenances for introduction will be guided by a broad knowledge of the types of wood required, and of the ecology and plant associations of desirable species growing in donor regions of similar climate to that of the recipient region. The concept of matching climates should not be applied too rigidly, nor should all species of minor significance in donor regions l e ignored. Details of the choice and sampling of species for introduction are discussed by several workers.

Short-term studies of variation in adaptive traits in wild populations and in progeny grown in a nursery, greenhouse or phytotron can help to limit the material for field study and to select test sites. Examples of traits affecting growth and yield that are amenable to such studies are growth rhythm, hardiness to frost and drought, and disease and insect resistance.

Multistage field testing, with replication in time and space, is essential to accommodate large numbers of species or provenances arid to sample environments adequately. There is much diversity of opinion regarding the desirable type and size of test plot, cultural treatments, experimental design, duration of the test, and assessment criteria. Decisions on these aspects must be made in accordance with the aims of the programme. Selected references can be recommended for guidance in planning field tests.

The need for international cooperation in genecological and yield studies of forest trees has been stressed many times, but much still remains to be done. The aims of this cooperation should be to provide adequate seed collections, to ensure the preservation of gene pools sampled, to determine patterns of variation and their causes, and to formulate guidelines for advantageous transfers of seed. Exemplary 'work on rules for seed movement has been done for *Pinus sylvestris* L. in Sweden, several exotic conifers in Britain, *P. caribaea* in Queensland and *P. taeda* in the United States. Analysis of variation by means of multiple regression, as described by Morgenstern and Roche, is a promising new approach to the delineation of seed zones.

By determining and using the best-adapted, high-yielding races for each major site, costly errors can be avoided and this essential basis for achieving maximum genetic gain through population improvement can be established.

Yield Selection in Base Population

Individual Selection

Several methods of utilising selected individuals for the improvement of net economic yield in a base population will be considered in this section. They all depend on heritable variation in biological yield and net harvest index among individual trees.

Large differences of economic yield have been demonstrated at young ages among progeny lines and among clones. This means that individual selection can be effective for improving yield. Substantial genetic gains have been predicted for several breeding methods on the basis of known or assumed values of population parameters and various selection intensities.

Realised gain can be estimated when progeny of selected trees are compared to a random sample of the base population. Substantial gains have been realised in some progeny tests (Table 4). Details of young tests are important not only because they often forecast what will follow, but because rapid early growth can be of value *per se*, for example in quick suppression of weeds. even if it is not sustained.

Table 4. Genetic Gains in Yield realised through progressively more intense breeding methods

Species type stand	*Age expt. From seed, (years)*	*No of parents: families*	*Type plot*	*Percentage gains*		
				Height	*Diameter*	*Volume*
A. Phenotypic selection, wind pollination; intensity of selection not specified						
P. elliottii (P)	8	11:11	M	7,6	4,9	-
P. elliottii (P)	10	19:19	M	-	-	10,7
P. taeda (W)	6	31:31	M	3,6	5,9	13,0

Contd....

Contd....

Araucaria cunninghamii (P)	6,5	20:20	M	13,0	-	-
P. radiata (P)	12,5	23:23	M	-	10,0	-
B. Progeny-test selection, wind pollination; intensity of phenotypic selection not specified						
P. elliottii (P)	8	4:4	M	12,4	8,9	-
P. elliottii (P)	10	8:8	M	-	-	20,0
P. taeda (W)	6	14:14	M	6,2	10,3	-
Araucaria cunninghamii (P)	6,5	7:7	M	22	-	-
C. Phenotypic selection, polycrossing or partial single crossing; intensity of selection not specified						
P. elliottii (W)	6	18:18	M	4,6	-	-
P. elliottii (W)	6	19:19	M	10,1	-	-
P. elliottii (W)	6	9:9	L	-	-	[4] 39,4
P. elliottii (W)	6	9:9	L	-	-	[4] 38,0
P. elliottii (W)	6	15:15	L	4,3	6,1	[4] 11,8
P. elliottii (P)	11,2	9:8	M	-	-	17,9
P. elliottii (P)	13,5	6:8	M	-	-	22,3
P. elliottii (P)	15	10:12	M	-	7,9	20,1
P. radiata (P)	11	11:15	L	-	-	14,0
Araucaria cunninghamii (P)	6,5	9:12	M	20,0	-	-
P. radiata (P)	12,5	21:21	M, L	-	-	22,2
D. Progeny-test selection, polycrossing or partial single crossing; 30 or 50 percent of families saved						
P. elliottii (W)	6	9:9	M	[5] 8,5	-	-
P. elliottii (W)	6	8:8	L	-	-	[4,5] 21,3
P. elliottii (P)	11-15	9:9	M	-	-	27,5
E. Progeny-test selection of the best full-sib families; 10 percent of families saved						
P. elliottii (P)	11-15	5:3	M	-	-	32,5

The simplest breeding procedure is mass selection; that is, use of wind-pollinated seed from superior phenotypes. It can give worthwhile gains if heritabilities are high, but if heritabilities are low, selection on the basis of progeny test results is much more effective than simple phenotypic selection. When mating is restricted to selected parents, which is equivalent to conditions in a thoroughly isolated seed orchard, larger gains are obtained. After a further stage of selection is applied, in which only the parents having high general combining ability are used, greater gain is achieved. A small additional increase in gain, compared to the time and effort that would be required to obtain it, was possible with *P. elliottii* by choosing for production the best 10 percent of single crosses. These results are in general agreement with predictions of Shelbourne for *Pinus radiata* Don.

The strongest evidence of gains in yield through individual selection comes from some of the older progeny tests. The experiments involved separate, replicated, multiple-row plots of commercial stock and progenies of parent trees selected for superior phenotype. Comparisons were for volume or basal area production (Table 4) and basic density of the wood. The latest measurements were made at 12.5 to 15 years. By this time, tree canopies had been closed for about one half of the test periods. This means that most of the wood was produced under competitive conditions within progenies.

Ages of the trees were more than one third of the likely rotations of the two species, *P. radiata* and *P. elliottii*. This suggests that the results may be reliable indications of the relative yields near rotation age. Basal area growth curves for several full-sib progenies of selected trees and commercial stock of *P. elliottii* and photographs indicate the selected material has established a distinct superiority. Moreover, the basic density of the wood of selected and commercial progeny was the same. This means that gross economic yield, in terms of both volume and dry weight of wood, has been increased by selection.

Families from selected trees of *P. elliottii* and *P. radiata* in the progeny tests referred to above also show considerable improvement in stem straightness. This improvement of stem quality will result in a higher net harvest index for most end uses. It is concluded that an increased net economic yield of about 30 percent to age 15 years may be expected from the establishment and subsequent culling of some *P. elliottii* and *P. radiata* seed orchards in Australia. Breeding programmes of this type are well worthwhile, since recent studies show that gains of the order of only 5 percent at harvest time are needed to justify the costs.

Problems of Yield Improvement within a Population

Individuals selecting for high yield

Net economic yield of a given forest is obviously a very complex characteristic having many components (Table 1). Often, when a breeding programme has begun little is known about relative economic values, degrees of genetic control and interrelationships of the factors influencing yield. Choosing selection criteria and allocating relative weights to them must therefore be subjective at first. Usually a systematic schedule is developed at the beginning of a programme to aid the selection of superior phenotypes.

More efficient means than phenotypic selection are needed for identifying high-yielding genotypes. Progeny tests have often been established to compare the breeding values of parents, but the high cost is unlikely to be justified for this purpose alone. Selection within progeny tests can be made with increased accuracy by combining information on individual, family and parental performance. Such combined selection will be used widely in tree breeding where suitable progeny tests are available.

Selection indices are especially useful where several complex and correlated traits are to be improved simultaneously. The construction of multiple-trait indices for rating an individual tree on its own performance is described by Northcott, van Buijtenen, and Illy. Worthwhile indices must be reliable, that is, based on precise estimates of genetic, environmental and economic parameters, but this requires large and expensive experiments.

Integrating evaluation, selection and estimation experiments

Recurrent selection for multiple traits affecting yield requires the mating of many selected parents, since many progenies are needed to evaluate parents and their families, to measure realised genetic gain, to select superior trees for the next cycle of breeding, to estimate population parameters, and finally to produce an ordinary harvest. Mating systems and experimental designs which are efficient for all of these purposes should be employed in order to economise on effort.

Descriptions of several mating and experimental designs that are being used in tree population improvement have been published. Mating systems vary in complexity and cost from the simple and inexpensive wind pollination to the complicated and costly diallel patterns for controlled pollinations. The designs differ greatly in utility for the various purposes mentioned above. A brief criticism of mating designs is given by Libby *et al.* and aspects to be considered in establishing field tests are described by Stonecypher. The choice of mating and experimental designs should be made very carefully, having regard to the resources and objectives of the breeding programme and the biology of the species. The importance of using a systematic mating scheme and well-executed field designs in a breeding programme cannot be overemphasised. It is basic to the achievement of acceptable genetic improvement.

Importance of population parameters

Population parameters of prime importance to the breeder are genetic, environmental and interaction variances and covariances of specific traits. Although much progress has been made in improving forest trees without precise information on this knowledge should increase the efficiency and facilitate all aspects of the planning and execution of tree breeding programmes. Mention has been made already of the need for good estimates of genetic and economic parameters for the construction of reliable selection indices. The importance of genotype environmental interactions and the use of population parameters to choose between alternative breeding methods will be discussed in the following sections.

Published heritability estimates for yield and other traits were summarised by Hattemer, but the methods of estimation of many of the ratios were criticised by Namkoong *et al.* These and additional estimates, again based mainly on young trees, suggest that heritability values for wood density, tree height and diameter rank in that order. Stem straightness usually has a rather high heritability.

Knowledge of the magnitude and sign of genetic correlations among traits affecting yield is important in selection. For example, strong negative relationships among some characteristics may mean that simultaneous improvement of them is impossible, even if selection indices are used. Where strong positive correlations are known to exist for a pair of traits, selection can be concentrated on the one which is easier to assess.

Genotype- environment interaction

Genotype-environment interaction refers to dissimilar performance of two or more genetic entities when grown in different environments. Interactions may be large enough to warrant the selection of separate strains of trees for specific environments. The most important environmental factors appeared to be locality, site and cultural treatment. Cultural conditions known to induce important interactions include site-preparation, fertiliser, irrigation, stocking and competition. Short-term screening of progeny lines for superior response to cultural treatment may not be reliable, and extended field trials may be needed.

The most practical means of minimising losses due to cryptic interactions is to use a broadly based genetic population for regeneration. Until specific genotype environment interactions are known in a particular breeding programme, selection should preferably be made under environmental conditions similar to those in which the progeny will be grown. Breeding plans should provide for testing progeny in several environments representing the major sites and under cultural practices recommended for current and future commercial planting.

Selection of Breeding Method

A dynamic breeding programme should have two goals: early production of improved planting stock, and further gains in subsequent cycles. The various methods available for achieving these aims differ in intensity of selection, mating procedure, aids to selection, and the time until improved seed or rooted cuttings are produced. All affect the grain expected per year. Since the cost of different breeding procedures varies also, a basis is needed for choosing the strategy having the greatest benefits in relation to cost for a given situation.

If satisfactory estimates of population parameters and the expense of various procedures are available, the gains expected from the use of different breeding procedures can be predicted, and the methods can be compared for cost and benefit. Shelbourne recommended that the most practical method giving near-maximum gain per cycle should be used in tree breeding.

In choosing a breeding method, account must be taken of expected gain, biology of the species, human and physical resources available, scale of the reforestation scheme, and costs and benefits both in the initial and later stages of the programme. The expectation of rather large gains from efficient breeding programmes (Table 4) justifies intensive breeding work, but additional critical studies, like those of Davis, of the economics of numerous alternative breeding procedures are needed to guide the choice of breeding method for a range of situations.

Hybridisation

Hybridisation of forest trees means the crossing of genetically differing entities such as species, varieties, provenances, or selected strains. Rarely does it involve crossing inbred lines of cross-pollinating species because of production difficulties. Derivatives of F_1 hybrids (F_2s and backcrosses) are included in the broad term "hybrids."

Two broad objectives of hybridising forest trees are: to combine valuable traits of two or more populations, and to obtain heterosis or hybrid vigour.

Combinational hybridisation is especially useful -in breeding for pest resistance and other multiple traits and in developing trees for sites to which a pure species is not well adapted. The needle-spot-resistant *Thuja plicata* Donn. x *T. standishii* Carr. and the well-formed, easily rooted *Populus alba* L. x *P. grandidentata* Michx. are good examples. Several hybrids of *Pinus* promising combinations of superior yield with other desirable traits have been reported recently.

Heterosis is exhibited when the mean yield of a hybrid exceeds that of its parents. It depends on the difference in gene frequencies between the two populations that are crossed and the level of dominance of alleles at loci affecting the traits concerned. Hybrid vigour is often manifest in the cross *Larix leptolepis* Gord. x *L. decidua* Mill. In many poplar hybrids, and in numerous other interspecific combinations.

Interracial hybridisation has been attempted less frequently than crosses between species, with encouraging but inconclusive evidence of beneficial results. Gains through heterosis and combinational effects may be expected, however, especially in species that have extended ranges covering a great diversity of habitats.

Sometimes backcross progenies are more promising than F_1 hybrids because of good performance and greater ease in the production of commercial seed. Advanced-generation hybrids may be expected to form the basis of future improved strains of forest trees.

Realisable Gains

F_1 hybrids often perform very well over a wide range of environments, probably because of their highly heterozygous genetic composition. But, when gains are properly measured in relation to the performance of the best alternative pure species available for a given site, it is often found that the occurrence of hybrid superiority is limited to environments in which neither parent is well adapted for high yield. These two points are illustrated by the fact that nine-year-old F_1 hybrids of *P. elliottii* and *P. caribaea* grow very well on both ridge and swamp sites at Beerwah, Queensland, but their superiority to the best parental species on the unfavourable swamp site is much greater than on the ridge site, where both parental species thrive.

Reliable quantitative estimates of gains realisable through hybridisation are rare, either because satisfactory controls are not always included in tests of hybrids or the progenies are still too young for evaluation. Nevertheless, there are strong indications that F_1 hybrids or their derivatives may yield worthwhile gains under some circumstances. For example the nine-year old F_1 hybrids of *P. elliottii* and *P. caribaea* described by Slee exceed *P. caribaea* by 21 percent in height and 20 percent in girth (and exceed comparable *P. elliottii* by an even greater margin) when grown on a swampy site.

The likely gain in net economic yield of the hybrids at age 11 years will be enhanced by a higher net harvest index due to better stem quality. Recent results from seven-year-old studies suggest that some backcross progenies grown on swampy sites may give gains at least equal to those of the F_1s. Since backcrosses are easier to produce than F1s, they are more likely to be used commercially.

Problems of Hybridisation

There are several reports of variation in hybrid performance according to the geographic source of the parent populations and the individual trees mated. Phenotypic selection among and within populations being hybridised may be effective for traits with high heritabilities, but extensive screening for parental combining ability is desirable for traits having little additive genetic variance. Performance testing and allocation of hybrids to suitable sites are problems requiring long-term observations.

Mass production of hybrid trees is simple for those in which vegetative propagation can be accomplished readily. But problems of parental flowering times, low seedset, and poor viability of F_1 hybrid seed usually combine to make mass production of seed very expensive. While backcross and F_2 hybrid seed is often highly fertile and more easily produced by natural pollination, the performance of these hybrid derivatives must he tested widely to prove their superiority before their commercial use is warranted.

Since hybrid planting stock is almost always more expensive than that of a pure species, a hybrid needs to be capable of rather larger increases in economic yield than those obtainable through population improvement. It is of the greatest benefit when it can be mass produced economically, when it is higher yielding than the parent which it is a candidate to replace, and when it is superior to any alternative pure species. In view of these several requirements of hybrids, they have rather infrequent, specialised uses.

There is growing awareness of the separate contributions that can be made to yield improvement by breeding, intensive site preparation, fertilisation, weed control, attention to initial spacing and periodic thinnings, protection, utilisation standards, and planned exploitation. But the greatest returns from investment in attempts to improve quantity and quality of forest economic yield must come from an integrated programme employing genetic, environmental and managerial approaches. A combination of these several approaches offers the greatest opportunity for increasing the rate of forest production.

4

Breeding for Stem Quality

Stem quality is defined by external or morphological properties as distinguished from the wood quality defined by internal or cellular properties. Thus, stem quality is mainly characterised by stem form and by growth and development of the branches. A tree may be classified as good or bad depending on the purposes for which it is intended and on demands as to quality. For saw-timber production, a tall, straight, healthy tree with fine, flat-angled limbs, a narrow crown, and a well pruned stem with 1 little taper is considered a good tree. The yield and quality of timber will be satisfying and the costs for handling, barking and transport will be relatively small. The same demands on the tree are valid for pulp production, but requirements as to straightness, taper and pruning might be less exacting. For trees intended for shelterbelts or recreation areas the demands will be different.

Breeding for improved stem quality will be restricted to the stem and branch characteristics important to quality.

Breeding Methods

Provenance tests have revealed a great variation of stem

and branching characteristics due to geographical sources. A careful test of foreign provenances and exotics is recommended in advance of transfer to and planting in unusual environments.

Progeny tests with half-sib, full-sib, and selfed progenies have given valuable information on variation of stem and branch properties among natural populations, stands, progenies, and individual trees.

The patterns of inheritance and the degree of inheritance of continuously varying characters could be studied more intensively and the variation of the characteristics assessed with greater reliability, once properly designed experiments had been established.

Improved techniques for measuring have been developed, and statistical methods for estimating genetic parameters in forest tree populations have been improved. Estimates of potential gains from breeding and of heritability made during the period up to 1963 have been summarised and discussed by Schreiner and Hattemer.

Studies of stem and branch properties, and their effects on wood quality, are reported for many of the most important conifers and even for some hardwoods.

The influence of silvicultural and other environmental factors on the variation of stem and branching characteristics is emphasised in all investigations. The degree of variance due to these nongenetic factors changes consistently with the traits investigated.

Characters for Stem Quality

Stem Properties

Straightness

Straight stem is generally considered to be a desirable

trait for all purposes. Deviations from straightness, such as lean, crook, sweep and twist, reduce the value and volume of the commercially useful part of the stem and may increase the costs for handling and transport. Methods used for measuring and grading defects are either subjective ratings of stem straightness or counting the number of crooks per unit length and measuring the deviation of the most severe crook or crooks. Photographs of a tree from different angles or a photogrammetric technique have also been used for quantifying bole straightness.

Stem taper

Uniformity of growth and good stem form, as expressed by the ratio between stem diameter at breast height and at a higher level, are important for volume production and properties such as wane and spiral grain.

Genetically controlled growth anomalies

Forking and fasciation in annual shoots disturb normal growth and cause crookedness of stem, development of two or more stems, and heavy branches. In the most severe cases the whole tree is twisted and bushy. Lammas shoots also give rise to crooked sterns and large branches which compete with the terminal leader and result in the formation of forked boles. The leader may retain dominance in later years, in which case the lemmas shoots produce strong branches, or it may be weakened by the competition and pushed aside, which may result in crookedness of the stem and in transforming the leader into a large branch. Ramicorn or fastigiate branches are sometimes formed repeatedly in an individual or clone.

Pruning ability

The age at which the lower branches of a tree begin to die

and fall off and the capacity of the tree rapidly to occlude scars and stumps are of great importance. After artificial pruning is carried out, both the speed of occlusion and the time to regain normal growth rate are essential features.

Branch Properties

Branch diameter

Branches small in diameter will leave smaller knots and cause less loss of wood quality. The risk of damage by environmental factors such as snow, ice, and attacks by fungi and insects will be reduced in a fine-limbed tree. The length of the branch is usually closely correlated with its diameter. Hence, crown width and the relationship between crown width and tree height are related to branch diameter. Ocular estimates or more precise measurements have been used for grading branch size.

Branch angle

The importance of flat or wide angles of the branches for stem and wood quality is weld recognised. Branches with flat angles are more easily shed and more rapidly occluded. Trees with such branches have less knot volume than trees with steeper branch angles. Ramicorns usually have very steep angles, an unfavourable trait. The risk of damage by environmental factors of the same kind as those mentioned above is increased when the angles are acute. The angle of the branch from the centre dine of the stem usually is measured with a protractor or ocularly classified as acute, intermediate or flat.

Number of branches, number of whorls, and internode lengths

These properties are essential features with great influence on stem development and wood formation.

Uniformity of growth, growth rate, straightness of stem, and knottiness are affected by the number of branches per whorl, the number of whorls per year and per unit length of stem, and the regular development of branches.

Wood Properties

Knotwood

Volume of branch knots has been calculated by measuring knot diameter, knot angle and knot length from the pith to the outside of the tree. This feature is closely related to the formation of compression wood and spiral grain, and is one of the main determinants of quality of the wood.

Compression wood

This property is characterised by abnormal tracheid development, large intercellular spaces and an unusual amount of shrinkage. The term compression wood is applied in the case of softwoods, while the term tension wood is used in the case of hardwoods. Reaction wood, a common term for both cases, is defined by Dadswell as "applicable to the wood formed by hardwoods and softwoods, which is anatomically different from normal wood" and it "usually involves eccentric radial growth of the stem, which in hardwoods is on the upper side of leaning stems and branches, whereas in softwoods it is on the lower side."

Grain patterns

Spiral grain or grain distortions generally cause reduced strength and loss of value of a tree stem. Twists will tend to occur in material containing spiral grain. Some unique grain distortions that create attractive patterns upon the face of lumber or veneer can have extremely high values.

Tracheid length

Cell length, cell width, and cell wall thickness each affect the yield and quality of wood and influence the specific gravity.

Specific gravity

This is widely used as a criterion of wood quality and yield. It is an excellent index of the solid substance contained in a piece of dry wood. Specific gravity values are obtained from cross sections of a tree or from increment cores.

Relations between stem and wood quality

Straightness of stem is essential for the formation of good quality wood. Crooks, sweeps and twists initiate the development of compression wood and irregular grain. The amount of severe compression wood is large in trees with severe crookedness and sweep. As the intensity of compression wood increases, tracheid length and specific gravity decrease, causing substantial losses in volume and value of a tree stem. Forking and ramicorns negatively influence the anatomy and structure of wood.

Size and number of knots, and the compression wood and grain pattern associated with knots, are mainly a result of branching characteristics such as large diameter, steep angle, numerous branches per whorl and numerous whorls. Indirectly, natural pruning ability belongs here. Early self-pruning reduces the knotwood volume in a stem since branches will be shed when comparatively thin, scars will heal more rapidly, and growth of the annual rings will regain regularity sooner. Also, pruning height, which is strongly related to total height of the trees, is important to the length of the merchantable portion of a tree stem.

Stem taper affects not only total volume production but also some properties of the sawn timber. Structure and strength of the wood especially are influenced. In trees with large taper, inclined grain is more pronounced than in trees with little taper.

Genetic Control of Characteristics

Patterns of Inheritance

Indications or proof of the occurrence of simple mendelian inheritance of stem and branch characteristics are scarce. Deformities ascribed to the effect of one or a few recessive or dominant genes are reported by Schröck and Ehrenberg. Forking and fasciation of annual shoots in Pinus sylvestris L. were observed in progenies obtained after open pollination or after controlled crossings among parent trees, which displayed the same abnormal growth. The production of lemmas shoots which cause forks of a different nature is a genetically controlled feature which probably has a more complicated genetic background.

There are some indications of a simple mode of inheritance of the form *virgata* in *Picea abies*. This snake form, at its extreme, completely lacks branches. Intermediate forms are found ranging from the extreme type to normally branched trees. Recessive genes, or a major gene causing bud and branch reduction and modified by minor genes and environmental factors, could be the explanation for this.

Stem characters such as straightness and pruning ability may be associated with sexual dimorphism, as is indicated in *Fraxinus* spp. and in poplars. If this is true, a 1:1 ratio of these characteristics could be expected. In *Populus nigra L.* the straight stem form is inherited as a dominant character and depends on more than two loci.

The results obtained from studies of artificially produced hybrids between *Populus nigra, P. deltoides* and *P. trichocarpa* indicate that, if the development of side branches in one-year-old seedlings is due to only one gene, which has been assumed, at least three alleles of this gene must exist. In the same material incomplete dominance of stem straightness was found and the development of an open, closed or intermediate crown form could be assigned to the effect of a small number of major genes. The influence of environmental factors on properties of this kind is great, however, and the length of time needed to obtain the F_1 and F_2 generations makes it a difficult task to determine the mode of inheritance.

In general, stem and branch characteristics are continuously varying traits controlled by quantitative inheritance. Additive genetic variance has been reported in all cases in which a statistical analysis of the material was possible as a result of experimental designs and methods used for assessing variation of a character. Evidence of dominant gene action (specific combining ability) is found in some cases, as for straightness and volume. Estimates of the phenotypic and genotypic correlation of one stem character with another have been made in many species. These help to clarify inheritance patterns. However, when genetic variances are being estimated, many assumptions have to be made which are never fully realised in the material investigated. This fact lends a varying degree of unreliability to calculated values.

Heritability

Estimates of heritabilities have accumulated during recent years. Here the same varying degree of unreliability of estimates holds as for estimates of genetic variances. A review of the published estimates cannot be made here. Certain trends in the material will be mentioned.

The range of the estimates of heritability is wide in every characteristic investigated. It depends on experimental designs, species, age of trees, environment, and other factors.

Most estimates are obtained from young or very young trees, usually seedlings and clones from one to ten years old. Only a few of the plantations 10 to 20 years old have been evaluated. In general, estimates have been obtained from conifers, especially fast growing pine species in subtropical regions, and from poplars.

New Breeding Methods and Techniques

Many of the methods and techniques developed in recent years are fundamental and common to several research sections of forest tree breeding. An attempt to summarise the development of and improvements in the most important sections will be made here, and some techniques concerned particularly with stem quality will be mentioned.

Field experiments, specifically designed for limited and well-defined purposes, have been established for progeny testing and for provenance and population studies during the past decade. Measurements and observations from these young plantations are now being published. Valuable information is being obtained on genetic variances, on correlations between characteristics, and on genotype environment interactions.

The heritability study established with Pinus taeda by the North Carolina State Cooperative Tree Improvement Programme, the International Paper Company and others, in Southlands experiment forest, Georgia, may be mentioned as such a long-term basic research project. It is especially designed for studies on quantitative genetics. Mating designs and pollination techniques have been developed and improved for population studies and for progeny testing of seed orchards.

Vegetative methods of propagation have been improved for many species, and new techniques for nursery management have been developed in order to facilitate the establishment of clonal seed orchards and clonal tests. These are essential parts of most tree breeding programmes.

Improvements in the efficiency of selection and in procedures for estimating genetic gains have been discussed by many authors. Recommendations have been made according to the results obtained and summarised. New methods and techniques have been developed for assessing stem quality. A photogrammetric technique for measuring bole straightness with high precision has been developed by Shelbourne and Namkoong and applied by Shelbourne in his study on the inheritance and relationships of bole straightness and compression wood in southern pines.

Much of the improvement in methods and technique is due to the rapid development of statistical methods and to the possibility of using computers for evaluating data. The background and details of these methods may be found in standard textbooks.

Finally, the establishment of phytotrons makes it possible to control environmental factors to a large extent. The assessment of growth rate, stem and branch characteristics, and wood properties at an early age will be feasible. Comparisons will be possible, not only among seedlings and clones studied under controlled environments, but also with the same material grown under different natural environments and at different ages.

Results with Different Species

Recent research into stem characteristics has mainly concerned some fast-growing conifer species of high economic value. However. information on slower growing softwoods in the temperate regions and on hardwoods in all regions, but especially tropical and subtropical areas, is accumulating. Information on variation in plantations and natural stands and the degree of inheritance of stem properties is particularly needed.

Some results recorded from experiments specifically designed for studies on genetic variation, phenotypic and genotypic correlations and heritabilities may illustrate the kind and amount of information which has been obtained in such plantations. Assessments were made on randomly selected parent trees and young progenies from one to seven years old.

In *Pinus taeda L.,* one of the most intensively studied species in this respect, a moderate heritability of straightness was demonstrated. Inheritance seemed to be relatively high for crook, but sweep was moderately to weakly inherited. Highly significant differences between progenies were indicated in progeny tests with open pollinated and control-pollinated progenies. High genetic components of variance were suggested for these traits. A moderate, positive parent-offspring correlation and also a moderately strong juvenile-mature relationship were demonstrated.

The same degree of inheritance of stem straightness was indicated in progeny and clonal plantations of Pinus elliottii Engelm., *P. palustris Mill.*, and *P. radiata D. Don.* A high general and specific combining ability for crookedness was evident in some cases. Similar evidence for a slight or moderately strong inheritance of stem features is indicated in other conifers, too.

Stem taper and ability to self-prune are genetically controlled features. According to Georgi and Schröck, significant differences in taper occurred between one-parent progenies of different origins in a 20-year-old test plantation of *P. sylvestris L. Barber* estimated the heritability for inherent pruning ability ranging from 0.50 to 0.64 in a progeny test of Pinus elliottii at eight years of age. He recorded a low but significant correlation between the height of natural pruning and both the second and third-year height growth ($r = 0.18$ and 0.16 respectively). In a provenance test with *Larix decidua L.* great differences between provenances indicated a fairly strong inheritance of the ability to self prune.

It is well known that branching characteristics are very much influenced by environmental factors, as are growth and other morphological traits. However, there is much evidence for genetic control of these properties too.

Branch diameter, which is closely correlated with branch length, varies with tree diameter, but heritability estimates indicate a moderate to weak inheritance. Branch angle is more strongly genetically controlled. Highly significant differences between progenies are recorded in progeny plantations and clonal tests, and a great part of the variation in this character is ascribed to genetic factors.

The tendency to develop ramicorns or very acute angled large branches is supposed to be genetically controlled. The number of branches per whorl and the number of whorls are fairly weakly inherited, but quite strong genetic control is reported from very young material of *Pseudotsuga menziesii.* Correlations between branching characteristics are reported in most cases studied. For instance, negative relationships were shown between branch angle and size, and number of branches per whorl and branch size.

Recent research in hardwoods shows great variation in each characteristic studied. High estimates of heritability for number of branches were shown for Populus deltoides. Significant differences between materials in respect of stem straightness, branching characteristics and natural pruning were recorded in Populus species by Einspahr and Benson and Benson and Einspahr and in Bombax ceiba L. by Venkatesch.

Highly significant differences in branch angle size among trees within stands and significant differences among stands within areas were reported by Kellison in Liriodendron tulipifera. Epicormic branching was reported to be a problem in Liquidambar styraciflua. This trait was mainly influenced by environmental factors, but there were some indications that- it might be genetically controlled too.

In *Dalbergia sissoo Roxb.* stem straightness is reported to be under considerable genetic control. There is a growing interest in hardwoods in the southern United States. A large research programme has been elaborated for studies of the species to be used in forestry. Investigations on geographical variations in some of the species have been made, and trees have been selected for future use in the breeding programme. Eucalyptus is a large and complex genus now generally planted on large areas outside its natural range. Inter- and intraspecific variations of growth, volume production, wood properties, phenological traits, and adaptation of the most valuable species are being studied, but little is known about variation and inheritance of the features.

Most of the earlier experiments were not properly designed for estimating phenotypical and genotypical correlations, variance, and heritabilities. Neither did they render reliable information on the variation of the stem and branch characteristics in wild or in plantations. However, results obtained from earlier studies are mainly confirmed by recent results in mature stands and in young test plantations.

Early tests have given valuable information about the later development of stem quality. Results of such tests are reported, for instance, in young plantations of *Pinus sylvestris* and *Pinus taeda.* A fair correlation of stem and branching characteristics in young seedlings with the same traits in the trees some years later was revealed. This indicates that it may be possible to evaluate mature stem features at an early age.

Genotype—Environment Interaction

The influence of silvicultural factors on stem and branching characteristics was long ago recognised and documented. Site, cultivation, spacing, thinning, fertilising and artificial pruning all affect stem quality. Changes of these factors add conspicuously to the variation of the characteristics. Change with site and climate is well demonstrated in numerous provenance tests and in progeny tests at different localities. However, test plantations specifically designed for evaluating the response of individual clones or progenies to changes in space and soil conditions are rare. Some information may be obtained from seed orchards and clonal tests, where the same clones are included in several plantations.

In a study of clones of *Pinus sylvestris* planted on rich clay soil and on poor sandy soil at three different localities in southern Sweden, the expected variation between clones was exhibited in each test. The individual clones changed with site with respect to branch characteristics, but the ranking of the clones in these features seemed to be about the same in all three test plantations. Clones with inherently narrow crowns remained relatively narrow crowned. The response to the changing environment went in the same direction for all genotypes investigated. The branch angle was not affected by differences in soil and climate, as was shown by the nonvarying clonal means. According to Keiding and Olsen, who studied two clones of Larix leptolepis planted on clayey and sandy soils at two localities, a fairly weak influence of the environmental factors on stem form was. revealed, and a very weak ortet-ramet relationship was demonstrated.

In a clonal test with Pinus sylvestris, designed specifically for testing the genotypic response to different spacing and fertilisation, a similar reaction in branch angle of four clones was indicated. The same difference between clones in branch angle remained even after different spacing and fertilisation treatments. The trees were young, four to five years from grafting, and were planted at two different spacings. Gansel found some effects of irrigation on the formation of crooks in slash pine.

In a six- to eight-year-old clonal plantation the trees in the irrigated plots showed an increased number of crooks per foot and a higher index of degree of crook compared to those on the nonirrigated plots. The results referred to above illustrate the reactions to be expected. Besides various clonal responses, even individual characteristics within a tree may differ in their response to the varying environment. The genotype-environment interactions are complicated, and much more knowledge of their nature and of the methods and techniques needed to obtain the necessary information is required. New silvicultural methods will continuously be tested and used in forestry, and must be considered when improvement programmes are being drawn up.

Future Approach

Considering the importance of improved stem and branch characteristics for the production of improved wood quality, weight should be given to these properties in breeding programmes and in studies of forest tree genetics.

The inheritance of stem and branch characters has been shown to be strong, moderately strong or weak, depending on which property is studied, but genetic improvement seems to be feasible in most.

The influence of environmental factors and the effect of silvicultural treatments have been shown to vary with characteristics, with age, and with site, etc. More research is needed in this field to improve the evaluation of promising exotics and provenances.

The need for information on genotype-environment interaction has been intensively felt in recent years, when comprehensive programmes have been designed and carried out all over the world for testing and use of exotics and nonindigenous provenances. For instance, information and research projects are badly needed on Eucalyptus species now widely planted outside their natural range. The same is true for many American conifers. When transferred to other geographical areas, these species compete more or less successfully with native species on good sites or replace indigenous species in areas of unsatisfactory growth and development. In forestry, the trend of introducing exotic species or provenances is not new, but demands and needs for new material have greatly increased and accelerated recently. Problems associated with this part of forest tree breeding should be given high priority.

Closely connected with the choice of proper species and provenances is the selection of superior trees in wild populations and in plantations. The genetics of quantitative characters and population genetics should be intensively studied, as they are essential as a basis for selection.

When choosing the characteristics for which to breed, consideration should be given to the relative degree of inheritance of a character. Results obtained from clonal and progeny testing indicate that improvement can be more easily obtained in stem straightness and branch angle than in, for instance, branch diameter, which is more influenced by changes in the environment. Thus, a higher selection gain will be expected if the characteristics to be improved are strongly genetically controlled.

Clonal tests specifically designed for study of the effects of silvicultural treatments should be planted on a large scale. Progeny tests specifically planned for estimation of inheritance of stem quality and of genetic correlations between the desirable features should be established. In both cases the test plantations should be planted as near as possible to the parent trees. The validity of the estimates of the parent-progeny relationships will be increased by reducing the differences in growing conditions between the mature trees and the young progenies.

Conservation of valuable stands and trees and exploration of unknown forests are of great importance for selection purposes, for hybridisation, and for preventing loss of genetic variability, which is the basis for selection. Methods used or suggested for the conservation of genes are manifold and in most cases they are expensive or difficult to handle. Natural stands could be protected in national parks or in reservation areas, and new stands could be raised intermittently from seeds of the trees included, and planted in the same areas.

Individual trees could be propagated vegetatively and grown in clone banks or in botanical gardens. There is also the possibility of freezing seeds or pollen and storing them in seed or pollen banks. Results from tissue cultures indicate that plant tissues can be kept growing on artificial substrates for long periods and can even develop into normal plants under such conditions.

5

Breeding to Improve Quality

The term wood quality will be used to refer to the properties of clearwood only. The causes and consequences of defects such as knots, reaction wood, grain deviations and most other forms of xylem discontinuity belong logically to the section of this consultation dealing with stem quality. Nevertheless, it is necessary to stress that superior wood quality can often become fully effective only in the presence of superior stem quality.

For example, the natural variability within many softwoods results in the strongest piece of defect-free wood produced by a particular species being perhaps two or three times as strong as the weakest piece; yet, the presence of defects can reduce the strength of either by a factor of 10 or more. A large cluster of encased knots or severe sloping grain in a critical region can render wood of the highest intrinsic quality mechanically useless. Consequently, any prediction of the gains to be derived from improved wood quality will, for most uses, require the added proviso, other circumstances being equal. The most important circumstances will be those affected by stem quality.

As stem characteristics have proved deserving of and amenable to improvement by breeding, it is probable that they will frequently be accorded priority over wood quality in breeding programmes. On the other hand many wood products, especially board and paper products, are very sensitive to variations in wood quality, and no opportunity for improvement, should be overlooked.

Feasibility and Applications

For most of the wood properties examined so far, the results carry conviction of the role of genetics from the cumulative evidence rather than from the strength of results of individual experiments. Nevertheless, some general trends are beginning to emerge. and a number of specific results do hold considerable promise for future gains. It will be convenient to discuss each wood property separately. First the applications of each property will be outlined and then the available evidence for improvement of each property by breeding will be examined.

Wood Density

This property is closely correlated with major strength properties of wood, with pulping yields, and with pulp and paper quality, as well as with machining, gluing and finishing characteristics. For these reasons, wood density is the property that has been most widely studied genetically, even though it is a complex property dependent on a wide variety of anatomical variables. Despite this complexity, fairly large narrow- and broad-sense heritabilities have been reported for coniferous trees 5 to 15 years old. Heritability values for corewood (1 to 5 growth layers from the pith) are usually lower.

Relatively little work has been done on hardwoods, and-results so far are less promising. Broad-sense heritabilities of 0.2 to 0.4 (for example, for *Populus spp.*) are typical. Some reports of heritabilities much higher than this imply that much more work is justified before the possibility of improving hardwood densities can be dismissed.

Fibre Properties

For convenience the term fibre will be used to refer to true fibres of hardwoods and also to softwood tracheids. Variations in fibre properties are the main determinants of variations in wood quality. The proportion of fibres to other types of cells in hardwoods and their dimensions in hardwoods and softwoods influence wood density, and they also largely determine the quality of pulp and paper products.

Fibre Length

Inheritance studies of fibre properties are still in the early stages, but it appears that the degree of genetic control of fibre length is at least as strong in conifers as that of wood density. Known inheritance values for hardwoods are few, but some rather high values have been assumed from related data. Though the range of values is limited, fibre length is such a limiting feature to utilisation of hardwoods that even modest gains would be welcome.

Fibre Diameter and Wall Thickness

Considerable variation has been recorded in these properties in conifers, but very few inheritance studies have been published. Results so far have provided good evidence of genetic control, but it does not appear to be as strong as control over fibre length. The strong heritability values obtained for wood density are largely dependent on these fibre characteristics. It therefore seems that the large experimental error that is possible in measuring diameter and wall thickness may have depressed estimates of their heritability.

Proportions of Fibres in Hardwoods

It is well known that the proportion of fibres to vessels and parenchyma cells in hardwood is subject to environmental control, but little information is available on the importance of genetic control except from clonal studies of Populus hybrids.

Spiral Grain

The importance attached to spiral grain varies widely from country to country. Attitudes probably depend on the severity of its occurrence in locally used species, on end uses, and on the climatic contrasts which focus attention on the dimensional instability of spirally grained timber. Where problems do arise from spiral. grain, they can be severe indeed.

Heritability has generally been assumed to be high. This is borne out by Nicholls et al. who found narrow-sense heritability to be 0.66 for corewood of Pinus radiata, with lower values outside the corewood. Spiral grain is most marked in the corewood of this species, but it is not normally an important feature of outer wood. On the other hand, Zobel et al. found genetic variation in P. taeda to be of a nonadditive type, so recurrent selection to reduce spirality would not be very effective in this species. They also noted that the incidence of spirality does not appear to be widespread or serious in P. taeda.

Chemical Properties

Extractives

The heritability of extractives contained in wood (as

distinct from chemicals obtained by tapping living trees) has mainly been examined in the softwoods used for pulp and paper manufacture. A high content of extractives can reduce yields, increase the consumption of chemicals in processing and may severely limit the usefulness of a timber for groundwood pulping. Though the content of extractives differs widely between trees, studies so far have not indicated strong heritability. The influence of extractives on wood utilisation is very extensive, and the possibility of modifying their effects by breeding deserves much more attention in future.

Heartwood formation

An assessment of heritability in heartwood formation, though of great importance for many wood uses, requires, in most species, older trees than are generally available as clonal or family assemblages. The only published results are those of Nicholls in which gross heritabilities of 0.36 are reported for heartwood formation in P. radiata.

Cellulose/lignin ratios

Cellulose content could be important in determining the yields of chemical pulps, but the few studies carried out so far indicate such a small degree of inheritance that selective breeding would give little improvement. The list of wood properties studied so far is by no means exhaustive. For example a recent paper by Zobel et al. has underlined the importance of moisture content in wood used for pulping. Variability of moisture content within a species may be primarily a function of variation in wood density and heartwood formation. Similarly, brightness is a critical feature of many softwood pulps, but fluctuations in brightness have yet to be related to variations in the raw material, even though intraspecific variation is quite widely regarded by pulping operators as the cause.

Perhaps the current situation can best be summed up by saying that many important wood properties can undoubtedly be improved by tree breeding. This improvement need not be given absolute priority. If necessary it may be claimed as a bonus in other improvement programmes. Thus, wood properties could be considered during final selection of parents initially chosen for other reasons, or during later additions to a programme.

An important point that emerges is the need to judge each species, or even each strain within a species, on its own merits. No generalisation can be assumed. The ultimate physiological bases of variations in wood properties to assume that, because a particular wood property does not prove amenable to improvement by breeding in one species, similar results will be obtained even in closely related species.

Consequently, the greatest need at the moment is for published results of soundly based experiments to examine genetic variances and also to measure actual genetic gains. Large-scale, adequately planned genetic experiments, along the lines indicated by Bannister and Stonecypher, are likely to be few in number. Therefore, for many years we shall have to rely on cumulative evidence from progeny tests aimed at modifying one or two properties of real importance.

Methods of Improving Wood Quality

Tree breeding is only one of several processes that can be employed to manipulate wood properties. One example will suffice to indicate the general role of each process.

Forest Management

Management practice regulates wood properties in many ways - through species selection, through changing growing conditions as determined by the sites selected for each species, and through many other decisions such as those regulating the ages of rotations.

Example

Long rotations can be used to increase the proportion of sawlogs furnished to pulp mills in integrated operations. In general, long rotations will produce a high proportion of outer wood. For most softwoods, the resulting pulp chips would have greater wood density, longer fibres, thicker cell walls, more heartwood and extractives.

Silviculture

Silviculture controls tree growth through initial spacing, thinning and site amelioration. By these means rates of growth can be manipulated within broad limits. In some species, wood properties such as density and percentage of latewood can be modified correspondingly.

Example

Softwoods grown in open stands tend to form less latewood than when grown under dense stand conditions.

Wood Selection at the Mill

In a small sawmill there may be little need for wood selection beyond an ability to sort logs by species and to optimise the cut from each. In large integrated mills producing sawn timber, veneers, particle board and/or pulp and paper products, wood selection can be very complicated.

Example

Pruned logs must be separated from unpruned logs. Logs from tops of trees may be chipped for chemical pulp or particle board, or they may be diverted to the groundwood mill. Sawmill slabs may be used to furnish a long - fibred admixture to chips from other sources. The variety of wood properties that becomes available when several species are used is limited mainly by the demands of the different end uses for sufficient material to meet the requirements of production. In this sense the various sections of an integrated operation become competitors and only advanced systems analysis and operations research can decide what is the most economical division of the available timber between them.

Put simply in this way, it might appear that so many other options are available to timber producers and users that wood improvement by breeding is an unnecessarily laborious and time - consuming process, and one, that is hardly worth the effort. This might be so if the economics of management and silviculture did not place severe restrictions on what can reasonably be done in the field and if many end uses did not demand the same sorts of high quality material. As it is, economic limitations usually prevail, and many processes that start off as waste users eventually demand wood of better quality to meet marketing competition.

Prospects for Improving Wood Quality by Breeding

The extent to which tree breeding is used to improve wood properties required for a particular product must almost always be decided by economic considerations. The data required for economic analysis are:

1. The need for improvement projected in terms of technological trends over the next one-and-a-half stand rotations - that is, looking ahead not less than 30 years, even for fast-growing conifers.
2. The degree of improvement which is required and feasible. For example, are wood fibres 3.4 mm long really superior to fibres 3.2 mm long, and what would be gained if we aimed at 4 mm ?
3. The cost of achieving such improvements by breeding. Visual selection for stem quality will almost always come first, followed by the more difficult laboratory evaluation and reselection for wood properties. If improvement in more than one property is sought, compromises must be made in selection intensity and projected gain for each. If desirable characteristics are negatively correlated, the cost of remedying losses in one property must be debited against the gain from the other.
4. The increased value that various degrees of improvement will confer on the product.
5. The costs and benefits of the breeding programme must then be balanced against the cost and benefits of alternative methods if any are applicable.

It must be conceded that the value of improvement may not always be directly assessable in monetary terms. For example, if quality improvements are required to meet competition, there may be no direct cash benefit, but the industry may fail if no improvements are made. However, this only emphasizes the need to assess wood quality improvement (by whatever means it is obtained) as part of the cost of production.

Much of the information required for economic analysis is now available or will become available as breeding programmes come to maturity. Unfortunately there is a dearth of published data on the values of qualitative improvements in various. products. This and the shortage of reliable estimates of genetic gain for many wood properties comprise the two most immediate needs for assessments of breeding programmes.

A pioneering attempt by Namkoong et al. evaluates the results of increasing wood density when *P. taeda* is used for chemical pulping and fully acknowledges limitations imposed by data that permit analysis only in terms of total yield of cellulose. They conclude that: "based on these data, which are admittedly inadequate, one could rationally say that the best breeding procedure would be to give specific gravity a negative weight or at best ignore it in breeding programmes." This controversial conclusion should stimulate others to provide data for similar analyses that will include the corresponding qualitative gains.

Because of the rapidly changing demands of technology, improvements of only fundamental wood properties are likely to prove worth while. For many species other selection criteria have priority; wood properties should be included primarily to ensure that they are not affected adversely. Alternatively, the main objective may be to reduce within-tree variability in some fundamental property; this would contribute to easier and more efficient harvesting and manufacture and could ultimately prove to be the greatest contribution of wood quality improvement. But properties fundamental to one industry may be of little concern to another.

The most clear-cut case for wood quality improvement would arise in an industry producing a single end product from a tree species of such outstanding merit that only vigorous stems of good form were produced and in which all wood properties but one fully satisfied the needs of the product. If the deficient property proved to be more economically corrected by selective breeding than by any other means, and if the cost could be justified by the increased value of the product, then a watertight case for improvement would be established.

It should be axiomatic that any change in management or silviculture or any change or diversification in end products must inevitably have repercussions on breeding strategy. Breeding programmes must be flexible and adaptable. These in turn will require flexible methods of assessment.

6

Seed Production of Organic Vegetables

The organic vegetable industry is flourishing due to consumers preference for organically grown produce over traditionally produced vegetables. As a result, an increase in the variety and selection of many vegetables in retail supermarkets and restaurants throughout the United States has occurred recently. With the new regulation requiring organic seed sources for organically labeled vegetables, many organic growers are searching for certified organic seed. Smaller seed companies have produced the majority of organically produced seed to date. The commercial seed sector is starting to provide a more diverse selection of cultivars.

Commercial vegetable seed production in the United States occurs mainly on the west coast in Oregon, Washington, Idaho, and California. The weather conditions in these areas have low humidity, excellent soils, and a warm, sunny, dry climate during summer which is favorable for organic seed production. These environmental conditions allow for a lower incidence of disease pest problems and a longer growing season, which are necessary for high quality seed production.

The Northeast and Midwest have also produced vegetable seeds in the past but now with the demand for organic seed increasing, seed producers in these regions have begun producing organic seed. These areas can be successful for producing seeds of many crops, even though the climate does not permit an extended growing season.

Selection of a seed crop that will thrive in the environmental conditions of the selected area is critical for achieving economic profit for the seed produced. Though Florida is not presently a prime area for seed production, okra, sweet corn, field corn, watermelon, and some other cucurbits were produced using traditional methods in the past. Commercial peanut seed production also continues to thrive in Florida. There are areas of Florida that have the potential to produce organic vegetable seeds, such as watermelon, muskmelon, and other cucurbits. Conditions in Florida are not conducive for growing longer season crops such as lettuce and carrots due to prolonged high temperatures and other environmental conditions.

Many organic growers want to grow open-pollinated as well as hybrid cultivars. To be considered a hybrid, the seed must be the offspring of two parents that differ in one or more heritable characteristics. For example, a green bean cultivar that has excellent flavor but is susceptible to a certain pest crossed with a different cultivar that is tolerant of the same pest, the two cultivars could be crossed and the new hybrid might have both excellent flavor and be resistant to the pest, provided the parents are homozygous. Seeds collected from a grown hybrid cultivar will not be true to type when replanted and thus can not be saved to plant next years crop.

One of the alternatives to hybrids is open-pollinated cultivars. These are produced by selfing or outcrossing to the same parental type. Other open-pollinated cultivars are referred to as heirloom cultivars because they have been grown for many decades. There is no standard length of time that a cultivar has to be grown in order to be considered an heirloom cultivar but generally 40 years is minimum. Since hybrid production requires obtaining the two inbred parents for crossing, they are mostly produced by commercial seed companies who own or develop the parental stocks. Thus, for smaller commercial seed companies and producers who grow their own organic seeds, only open pollinated cultivars are produced.

Conventional and Organic Seed Production

Vegetable crops are produced for two distinct markets, fresh and processed. Conventional cultural practices are fairly similar. Growing crops for seed requires a longer season since the crop must stay in the field twice as long as a conventional crop harvested for fruit and also for increased monitoring to ensure high seed quality and purity.

Standard production of vegetable seeds requires chemical herbicides, insecticides, fungicides, and fertilisers. An increase in the amount of chemical products used on seed crops may occur due to the length of time the crops remain in the field. As a result, plant diseases and insects have more time to attack the crop during seed maturation. With these factors playing a role in conventional seed production, the challenges for organic seed production are increased.

Organic farming practices began on small farms in the United States during the 1940s. As the diversity of crops labeled organic increased, there became a need for certification standards and regulations to ensure proper guidelines for growing organic crops. By the 1980s, the organic industry increased dramatically, resulting in the establishment of the Organic Certification Board which issued the Organic Foods Production Act (OFPA), identifying "organic" produce. New laws now stipulate that as of October 2002, all producers and handlers must be certified by a USDA certification agent to label or sell a crop as "organically grown."

Guidelines for Organic Seed Production

Land Selection

Good farming practices are especially important when growing vegetable crops for seed. Proper selection of land is critical to ensure maximum quality and economic yield. Crops should be rotated to reduce pest problems and any potential for seed contamination by open pollination with similar species types. The seed production field should not have known weed problems that are too difficult to control through organic means as weeds will compete significantly with the crop, resulting in seed yield and purity problems. Weed management is a major problem with organic crop production.

Land Preparation and Soil Fertilisation

Soil should be tilled to ensure a fine seed bed which is critical for germination, particularly with small seeded crops. The soil should have good water holding capacity to allow for uniform germination and continued vegetative growth. In Florida, the beds should be raised and shaped depending on rainfall. It is especially important to have uniform seedbeds because the seeds are often precision planted and uniform emergence and seedling development are required for optimum management. Proper irrigation is necessary to ensure growth during droughts. Drip irrigation is an effective way to deliver water to the crop without wetting vegetative plant parts thus reducing potential disease problems.

It is important that the fertility of the soil is improved when producing organically since chemical fertilisers can not be used. To ensure good fertility and fewer soil diseases, crop rotation, use of a cover crop, green manure crops, mulch, animal compost, and plant material compost can all be used. Floridas soils are generally sandy which provides good drainage and ease of tillage, yet nutrients are leached readily during heavy rains. Organic or natural fertilisers are generally made from bone meal, blood meal, horse, and chicken manure, mushroom compost, ash from sunflower seed hulls, and many other substances. Considerations and modifications must be made depending on the specific land selection and soil conditions. There are many commercial organic fertilisers available for organic crops with the major ones listed below.

— Composted manure (mushroom compost is mostly manure), or blood meal or feather meal, used to increase nitrogen content. Dried blood is also a useful nitrogen source when used with wood chip mulching, which is high in carbon, and low in nitrogen.

— Wood ashes, or potash, for increased potassium.

— Rock phosphate, often crushed rock that contains elevated levels of phosphate.

— Seaweed extract - not a fertiliser, but aids plant growth and resistance to pests and disease.

Inoculates used with bean and pea seeds provide beneficial fungi which work with the plant's roots to help them fix nitrogen from the air. Cover crops used between seasons as green manure may increase soil organic matter for the growing season.

Planting Techniques

Seeds are generally planted directly by drilling in the field or transplanting from a greenhouse grown seedling. These seeds should be organically produced and all transplants must be organically produced. The seeds must be planted in such a way that proper vegetative development occurs that will support fruit and seed development with proper spacing and depth in the bed. Vegetable seeds are greatly influenced by temperature at planting; pea, lettuce, radish, beets, onion, and spinach grow best at cool temperatures for optimum emergence. Other crops including squash, melon, tomato, pepper, and eggplant grow well at warmer soil temperatures. In general, the best germination temperature for all these crops is around 70°F.

Row spacing and plant densities must allow for maximum plant development of the flower and unrestricted access to inflorescences for pollinators which will lead to proper fruit and seed set. Proper spacing will also allow for improved air movement reducing pathogens and space for harvest operations at the end of the season.

Pollination of crops for seed is also affected by row spacing and planting densities; using larger spacing allows for better wind movement and more room for insect pollinators. Crops that are wind pollinated include sweet corn and spinach and those that are insect pollinated include carrot, onion, and broccoli.

Weed, Insect, and Fungal Management

Management of weeds and pests are critical to ensure that organically produced seeds have high yield and quality. It is common for insect pests after harvest of the primary host crop to move to another crop in a neighboring field. This is the case for the Lygus bug. It will feed on legume crops but once harvested, they will move to umbiliferous crops such as carrots. Other insects such as cabbage looper infest crops in the Brassica family and causes loss of vegetative parts effectively reducing seed yield and quality. Aphids are a problem, especially at harvest because they secrete "honeydew" on crops including tomato, squash, and pumpkin making harvesting difficult. They also spread diseases, especially virus, which is difficult to control conventionally and even more challenging for an organic grower.

Because the seed crop is in the field for a longer period of time, there are more opportunities for multiple pathogens to interact with a single crop. There are also cases where pathogens target the seed directly and affect the entire seed population. To control these pathogens organically is complex and requires proper growing conditions such as temperature and humidity to minimise the incidence of such pathogens as powdery mildew which thrives in high temperature and humidity. Seed producing fields should not be located near commercial vegetable fields. Eliminating weeds will help decrease insect habitats and pathogen spread from host plants likely to infect the seed crop.

Biological Pest Control

Another viable pest control method is through biological control. This practice is highly compatible with organic seed production, and it does not involve the use of chemicals. Biological control is defined as the decline of pest populations by natural enemies. Biological control utilises three sources of natural enemies that can be used to control harmful pests and reduce the use of organic pesticides. The first group includes parasitoid insects, which lay their eggs inside another insect. An example of a parasitoid insect is the wasp, Aphidius colemani which lays its eggs in aphid adults. The wasp matures inside the aphid egg, killing the aphid and continuing the wasps life cycle.

The second group of beneficial insects is predators, which eat other insects. A common predator is the lady beetle, which preys on insects and mites. The third group includes the weed feeders, which feed on weed populations. Insects such as a weevil, Hylobius transversovittatus feed on certain weeds including purple loosestrife. Pathogens such as fungi, viruses, bacteria, and nematodes are also capable of destroying weed populations and are called bioherbicides. Reducing weed populations is highly beneficial because often times during winter, pests live in weeds surrounding fields used for seed production. It is important to note that there are many beneficial insects that can be used to control a large number of pests.

Harvesting, Threshing and Drying

Vegetable seed crops are separated into two categories - those that produce wet-seeded fruits and those that are dry seeded. Wet-seeded crops include tomatoes, peppers, cucurbits and those that are dry seeded include the brassicas, legumes, and onions. Harvesting method depends on the type of seed being produced.

When harvesting dry-seeded crops, seed shattering must be prevented because seed harvest generally occurs after the crop reaches physiological maturity. To reduce shattering, the stalks of the plant need to be cut while still green and field dried allowing for uniform seed maturation. Once the seeds are mature, they can be machine or hand extracted. Observation may help to judge how the plant is maturing and for signs such as drying of florets or other plant parts. This can be complicated and it requires good knowledge of the crop but can lead to excellent seed quality, because the seeds will be harvested at peak maturity and size.

Determining when wet-seeded crops are mature for seed harvest depends primarily on fruit color. Peppers and tomatoes are harvested at full color because the percent germination is much higher in the mature red compared to the green fruit. In any case, wet-seeded fruits can be harvested, the pulp processed, and the seeds removed. The seeds must then be washed in water then dried either in the sun or by rotating in a large drum before oven drying. Modifications must be made with certain crops, such as melons, to remove the mucilaginous coating around the seed. When the seeds are removed from the pulp and outer coatings, they should be dried immediately to prevent fermentation.

Cleaning and Storage

Once seeds are harvested, threshed, and extracted, they should be evaluated to determine if further cleaning is required to remove any debris such as dirt or plant parts. Small seeds such as carrots are sorted and separated using a debearder machine to remove unwanted outer seed parts. All seeds should be single units and all stored according to their individual temperature/humidity requirements. Seeds are best stored at 50% relative humidity at 50°F.

Pelleting Techniques

The commercial seed industry often pellets seed, if the seed is irregularly shaped or small. Developments in seed technology have allowed for artificial seed pelleting materials to aid in regulation of water uptake by the seed which can influence germination rate. Seeds are pelleted with materials that coat with multiple layers of inert materials which shape and color the seed for better visualisation during seeding.

The drawback for organic growers is that the normal commercially available pelleted seeds have substances used to pellet that contain synthetic ingredients not permitted by the standards of the National Organic Program (NOP). According to the NOP Rules of the USDA National Organic Program, use of the USDA organic label or seal Subpart D describes the use of the USDA organic label or seal.

- For seed producers, much of the labeling information is not applicable.
- For example, products may be labeled as "100% organic," "organic," or "made with organic ingredients." Since raw seed is a single composition product, it will either be "100% organic" or not.
- Seed treatments must be approved by the National Organic Standards Board (NOSB) for use on organic seed. Prohibited substances (such as synthetic pesticides) may not be used on organic seed.
- Pelleting and film coating materials must also be approved. The only time a seed product may not qualify as "100% organic" is if a pelleting or film coating material contains a minor component(s) that is not organic. At that point a certifier would need to determine if it could be labeled "organic,"

indicating a product that contains (by weight or volume) not less than 95% organically produced or processed agricultural products, or "made with organic ingredients," indicating made with at least 70% organically produced ingredients.

7

Reproductive Plant Breeding

Some plants reproduce by self-fertilisation where pollen from a plant can fertilise the reproductive cell or ovule of the same plant. Other plants only allow cross-pollination where pollen from one plant can only fertilise a different plant. Asexual propagation can also occur in plants (e.g. runners from strawberry plants) which gives a new plant which is genetically identical to its parent plant. All these differences change the way plant breeders work.

Mode of Reproduction

The mode of reproduction of a crop determines its genetic composition, which, in turn, is the deciding factor to develop suitable breeding and selection methods. Knowledge of mode of reproduction is also essential for its artificial manipulation to breed improved types. Only those breeding and selection methods are suitable for a crop which does not interfere with its natural state or ensure the maintenance of such a state.

It is due to such reasons that imposition of self-fertilisation on cross-pollinating crops leads to drastic reduction in their performance. Likewise, it is practically impossible to maintain permanent heterozygosity in self-fertilising crops rendering the development of hybrids an unattractive breeding approach. Asexual propagation is another type of reproduction where any plant or part of it can be used for multiplication without even a slight genetic change from generation to generation. The methods of breeding and multiplication for such crop are thus entirely different than those of sexually reproducing crops.

Self Fertilising Crops

Certain restrictions caused the mechanisms for self fertilisation (partial and full self fertilisation) to develop in a number of plant species. Some of the reasons why a self fertilising method of reproduction is so effective are the efficacy of reproduction, as well as decreasing genetic variation and thus the fixation of highly adapted genotypes. Most of the loci get fixed at a high rate; this can be ascribed to the fact that with each generation of self fertilisation the rate of heterozygotes decreases by 50%.

Homozygosity will thus be obtained in 5-8 generations. The 3rd reason for the efficacy of self fertilisation is that in mixed stands of self and cross pollinating crops, the self fertilising plants can donate pollen to both plant types, where the cross fertilising plants are restricted concerning the contribution it can make to the population with regard to pollen donation. Almost no inbreeding depression occurs in self fertilising plants because the mode of reproduction allows natural selection to take place in wild populations of such plants.

Thus, the genetically non-superior or unstable plants are removed from the population at an early evolutionary stage. Populations derived from self pollination are sometimes not as evolutionary adaptable as with other reproductive methods, but are known to utilise specific ecological niches more effectively. Critical steps in the improvement of self fertilising crops are the choice of parents and the identification of the best plants in segregating generations.

The breeder should also have definite goals with the choice of parents. Self fertilising cultivars are easier to maintain, but this could lead to misuse of seed. Some of the agronomically important, self fertilising crops include wheat, rice, barley, dry beans, soy beans, peanuts, cotton, tomatoes, etc. The basic methods by which new genetic variation can be established in self fertilising crops are by means of introduction, selection and hybridisation.

A few different selection methods are applied in self fertilising crops, such as mass selection, single plant-, pedigree-, bulk population-, back cross-, recurrent-, F1 hybrid, as well as single seed descend (SSD) selection. In most breeding programmes a combination of these methods are applied. The different selection methods are described below:

Mass Selection

This method of selection depends mainly on selection of plants according to their phenotype and performance. The seed from selected plants are bulked for the next generation. This method is used to improve the overall population by positive or negative mass selection. Mass selection is only applied to a limited degree in self fertilising plants and is an effective method for the improvement of land races. This method of selection will only be effective for highly heritable traits. Some of the shortages of mass selection are that it is not possible to know whether selected plants are hetero- or homozygous.

Heterozygous plants will segregate in the next generation and it may thus necessitate repetition of phenotypic selection. Secondly, the influence that the environment has on growth of the plants includes development, phenotype and performance. It is not always clear whether the phenotypically superior plants are also genotypically superior and strong environmental differences may lead to low selection efficacy.

Single Plant Selection

A variety developed by this method will be more uniform than those developed by mass selection because all of the plants in such a variety will have the same genotype. The seed from selected plants are not added together but are kept apart and used to perform offspring tests. This is done to study the breeding behaviour of the selected plants.

The high uniformity in stand and performance has been stressed in the past, but the risk of highly specialised pathogens evolving is very high. More genetic variability could buffer the crop against such pathogens as well as stability of production under varied environmental conditions.

Selection for Pure Breeding Cultivars from Crosses

Crosses between varieties, germplasm introduction and breeding lines are made to create new gene combinations. In the generations to follow, superior gene combinations are selected and fixed in the homozygous state by means of self fertilisation and selection. These selections are tested extensively with the goal of releasing them for cultivation.

Pedigree Selection

Parental lines are crossed and selection of plants with

new gene combinations already takes place in the F2 generation (the generation of plants formed from crossing F1 hybrids). The offspring of selected populations in the generations to follow are repeatedly subjected to selection until genetic uniformity is reached. Records are kept of the origin of the selected individuals or lines. The amount of generations of single plant and line selections, as well as selection intensities, can be varied in practice according to the crop and availability of facilities. It is usually traits with high heritability that are quick and easy to measure that are concentrated on when using this method of selection.

One of the main objections against this method is that the genetic variation, available for selection of quantitative traits, are drastically decreased in later generations. Seed purification and multiplication is usually incorporated in one of the final generations of pedigree selection. This method is very labour intensive. Breeding efficiency is one of the goals with early generation testing. This is done by early identification of superior heterogeneous populations.

The early elimination of inferior populations and subsequent concentration of selection efforts within superior populations is assumed to result in increased efficiency. Accurate evaluation of heterogeneous populations is essential to the success of this method and assumes that transgressive segregants from inferior populations will not exceed selections from superior populations in performance.

Bulk Population Selection

With this method of selection the offspring from a crossing are planted at planting densities equal to commercial planting densities. During this period, which may include a number of generations, the level of homozygosity in the bulk population increases. This method is simple and cheap and involves less work than pedigree selection in the earlier generations. It is necessary to plant large populations to ensure that the best segregates are selected when selection starts.

Segregating generations are subjected to another single plant selection step. Fewer records are kept during earlier generations than with pedigree selection. This type of selection is especially carried out with crops which are usually planted at high planting densities, e.g. small grain crops.

Single Seed Selection

This method was introduced to keep as many F2 plants as possible intact and thus prevent the loss of variation of traits with low heritability in earlier generations. This method is also used to decrease the time that is needed to grow segregating generations. Because only one seed is harvested per plant, it is not necessary for optimal plant development and conditions can be manipulated so that 2-3 generations can be harvested per year.

This can cause the process from the starting to finishing time to decrease by 1-3 years. This method does not eliminate weak plants such as in other methods and there is also no provision for selection of superior plants in the F2 generation. Modification of this method is possible and record keeping is not necessary in early generations.

Doubled Haploid Method

Haploid plants may be produced by chromosome elimination in wide crosses, ovule culture or by anther culture. Anther culture, however, is mostly used because of its ability to produce haploid plants with much larger quantity compare with the other two methods. Stresses are usually necessary to alter the development pathways of microspores from producing pollen to forming haploid plants.

The chromosome numbers of the haploid plants are then doubled with the use of colchicine. Spontaneously doubled haploid plants, however, can also be produced directly from the three methods. Embryo rescue methods can be used to ensure that seed from these wide crosses or stigma culture plants survive and doesn't get aborted. This method has the potential of shortening genetic improvement cycles in comparison to pedigree or bulk methods.

Like the single seed selection method, early generations are not subject to selection, but most of the lines [illegible] during the land evaluation trials. This [illegible] sive and the most expensive [illegible] reases the amount of [illegible] thod to be successful, the [illegible] le.

Back Crossing

This is a type of repeated selection where a specific gene can be incorporated into otherwise superior cultivars. One of the parental varieties is highly productive and commercially successful but lacks a specific gene (e.g. disease resistance). This trait is usually present in the other parental variety. After each back cross, hybrid plants are identified with the gene under consideration and are back crossed again with the repeating parent. This technique is easy when traits are added which are easily inherited, dominant, and easily identified in hybrid plants.

If closely linked, unwanted genes are present with the useful genes, the unwanted genes could be transferred together and the offspring may actually be less productive. An advantage of the back crossing method is that extensive testing is not necessary. This method is used to create hybrid varieties in self fertilising crops and to establish male sterility in parental lines. Marker-assisted backcrossing is routinely applied in breeding programmes for gene introgression.

Traditional Selection Methods

A Relatively small gene pool is created with traditional selection methods, there is not a lot of opportunity for gene recombination and few opportunities for the breakage of linkage blocks. Breeders are looking to overcome these shortcomings by increasing the amount of crosses made, as well as to start testing for performance in earlier generations and to introduce systems for improvement of populations (e.g. recurrent selection). Other breeding strategies include multilines (a composite of genetically identical lines which have different genes for example stem rust resistance).

Lines that are basically genetically identical, except for a single gene, are called isogenic lines. Mixed varieties (composites) are also used and are less uniform than pure varieties should be revised in regular intervals to prevent shifts in the proportion of components. Recurrent selection is most often used in cross fertilising crops, but can be successfully used in self fertilisers only when a male sterility gene is present. Hybrid varieties are a very important method in self fertilising crops and have the following advantages:

- Heterosis
- Easier to create varieties with multiple resistance genes

— Attractive method for private breeders because of built-in mechanisms for protection of varieties.

Raising hybrid seed has been one of the major goals of horticultural and agricultural practice, because hybrid plants are more productive (due to hybrid vigour) and more uniform in quality than plants derived from self-pollination or random pollination. To raise hybrid seed, self-pollination and sib-pollination (pollination by a plant of the same hybrid) must be circumvented. One method is hand emasculation of the line used as female parent, which is then naturally cross-pollinated by pollen from the line serving as male parent and planted in an adjacent row.

However, this process is very labour intensive and invariably expensive. If the crop plants can be made self-incompatible by the introduction of the genes controlling self-incompatibility, then all seeds produced will be hybrids resulting from cross-pollination between two different lines. This would facilitate the production and increase the yield of hybrid seed and, at the same time, reduce the labour costs.

Cross-pollinated Crops Selection

The natural state of self-fertilising crops is homozygosity and genetic uniformity, whereas cross-fertilising crops are characterised by a high degree of heterozygosity. Plant species where normal mode of seed set is through a high degree of cross-pollination have characteristic reproductive features and population structure. Existence of self-sterility, self-incompatibility, imperfect flowers, and mechanical obstructions make the plant dependent upon foreign pollen for normal seed set.

Each plant receives a blend of pollen from a large number of individuals each having different genotypes.

Such populations are characterised by a high degree of heterozygosity with tremendous free and potential genetic variation, which is maintained in a steady state by free gene flow among individuals within the populations. It is inappropriate, and could be rather hazardous, to take one or a few individuals to investigate or improve these populations.

The enhanced fitness of heterozygotes over homozygotes of cross-pollinated crops has been manipulated in the form of two different breeding approaches namely, population improvement and hybrid breeding in such crops. In the development of hybrid varieties, the aim is to identify the most productive heterozygote from the population, which then is produced with the exclusion of other members of the population. In contrast, the population improvement envisages a stepwise elimination of deleterious and less productive alleles through repeated cycles of selective mating of genotypes that are more productive.

Population improvement is slow, steady and a long-term programme, whereas the production of hybrids is aimed to maximise the genetic gains in much less time. Both of these breeding approaches are complementary rather than mutually exclusive and are based on sound genetic theory. The different selection methods can be summarised as follows:

Mass Selection Method

It is the simplest, easiest and oldest method of selection where individual plants are selected based on their phenotypic performance, and bulk seed is used to produce the next generation. Mass selection proved to be quite effective in maize improvement at the initial stages but its efficacy especially for improvement of yield, soon came under severe criticism that culminated in the refinement of the method of mass selection. The selection after pollination does not provide any control over the pollen parent as result of which effective selection is limited only to female parents.

Recurrent Selection

This type of selection is a refined version of the mass selection procedure and differs as follows:

- — Visually selected individuals out of the base population undergo progeny testing
- — Individuals selected on basis of the progeny test data are crossed with each other in every possible way to produce seed to form the new base population.

Half-sib Selection

Selections are made based on progeny test performance instead of phenotypic appearance of the parental plants. Seed from selected half-sibs, which have been pollinated by random pollen from the population, is grown in unreplicated progeny rows for the purpose of selection. A part of the seed is planted to determine the yielding ability, or breeding value, for any character of each plant. The seed from the most productive rows or remnant seed from the outstanding half-sibs is bulked to complete one cycle of selection.

Full-sib Selection

A number of full-sib families, each produced by making crosses between the two plants from the base population are evaluated in replicated trials. A part of each full-sib family is saved for recombination. Based on evaluation the remnant seed of selected full-sib families is used to recombine the best families.

Selections with Test Cross Performance

The purpose of this type of selection is a slight deviation from the concept of intra-population improvement in the sense that the population is improved, not only for performance, but also with respect to combining ability with a specific reference population. It involves genetic modifications of the population with an aim on its better use for the exploitation of heterosis. It involves three steps:

- — Self-pollination and test crossing of individuals
- — Evaluation of test crosses in replicated trials
- — Recombination from selfed remnant seed of selected plants

Selfed Family Selection

The plants in the original base population are selfed to produce S1 progenies, which are evaluated in the next season in replicated multi-environmental trials to identify promising S1 families. The remnant S1 seed of such selected families is then recombined in the third season as a result of which one cycle is completed in three seasons. Hence, the units of selection and recombination are S1 progenies.

Asexually Propagated Crops Breeding

Asexual reproduction covers all those modes of multiplication of plants where normal gamete formation and fertilisation does not take place making these distinctly different from normal seed production crops. In the absence of sexual reproduction, the genetic composition of plant material being multiplied remains essentially the same as its source plant. Clones of mother plants can be made with the exact genetic composition of the mother plant.

Superior plants are selected and propagated vegetatively; the vegetative propagated offspring are used to develop stable varieties without any deterioration due to segregation of gene combinations. This unique characteristic of asexual reproduction helped to develop a number of cultivars of fruits and vegetables including grapes, apples, pears and peaches.

The selection in these crops is restricted to the material introduced from other sources, such as field plantations. The promising selections are tested in large scale trials which, if successful, can be multiplied and released for commercial cultivation. The improvement of asexually propagated plants through induced mutations has distinct advantages and limitations. Any vegetative propagule can be treated with mutagens and even a single desirable mutant or a part of a mutated propagule (chimera) can be multiplied as an improved type of the original variety.

Selection, in the case of asexual plants, can be defined as the selection of the best performing plant and the vegetative propagation thereof. Because plants are not totally genetically stable, it can be expected that deviations would occur through the years. Selection is thus an ongoing process where deviants are selected or removed from the selection programme. The main purpose of selection is to better the quality and yield of forthcoming plantations. Such as any breeder, the breeder has to have a good knowledge of the characteristics of the cultivar under consideration.

Different approaches can be followed in the selection process of asexual plants, such as mass selection and clone selection from clone blocks. In mass selection there are some factors that must be considered when selecting plants in a mother block, e.g. vineyard. Time of selection is a big factor, because you have to select when most of the characteristics of the plant are clearly showing.

With asexual perennials the best time is just before harvest. For the best results the selected plant must be evaluated during the next season, when growth-abnormalities, leave disfigurations and virus symptoms are best visualised. Mass selection is done annually on the same plant for a minimum of three years. A plant that does not conform to the requirements in any given year of the selection cycle is discarded from the programme.

Older plantations which were exposed to harsh growth conditions are seen as a preferred selection sources. The plants that grew under these circumstances and performed well are seen to have good genetic properties. In these older plantations natural selection took care of most poor performing plants.

Development of New Clone

The development and registration of new clones take place by means of local clone selection in old plantations, as well as the importation of high quality clones from abroad, for local evaluation. A clone is the vegetative offspring of one specific mother plant; it does not show any genetic, morphologic or physiologic deviations from the mother plant. Evaluation takes place with the different selected clones after selection. The dissimilar clones are compared to each other to determine their quality and resistance capabilities.

Breeding is not involved in clone selection; the clone cannot be bred for resistance of certain types of viruses, emphasis must be put on making sure that the clone material leaving the nursery are virus free. Techniques are developed to test the clones for any harmful viruses. Harmful viruses sometimes do not show in the preliminary evaluations. Phytosanitary development (virus detection and virus eradication) is thus performed in laboratories and greenhouses, parallel with field- and quality evaluation in field clone trials.

When a clone complies with the minimum quality and phytosanitary standards prescribed by the Plant Improvement Association (PIA), it is officially registered for certification and commercialisation. The PIA is an association that complies for all plant improvement, including grapes, apple, pear and peaches. Clones are made of cuttings from a field-grown mother plant. Due to bad management and infection from neighbouring plantations there is only a few virus free mother plants in selecting plantations.

The clone developers had to incorporate techniques such as tissue cultures and in vitro propagation to develop virus free clones from the limited mother material. The apical meristem is free of any harmful viruses. By using the apical meristem for tissue culture a virus free clone can be developed.

Multiplication

During the first phase of multiplication, a nucleus cutting from each candidate and/or registered clone is kept in a PIA approved insect free nucleus block green house. From here all future multiplication and evaluation will be done. During the second phase of multiplication, rootstocks and grafted cuttings are established in foundation blocks in insect free facilities and open field isolated areas, from where further evaluation are done. The scion and rootstock material from the second phase source are grafted and callused.

The grafted plants are then planted in isolated areas for the establishment of mother blocks. These blocks are used for multiplication purposes. The third phase is the establishment of scion mother blocks from the above mentioned source on farms of contracted collaborator producers in pre-selected virgin soil of which about 4 km^2 are maintained.

8

Plant Pollination

Pollination is an important step in the reproduction of seed plants: the transfer of pollen grains (male gametes) to the plant carpel, the structure that contains the ovule (female gamete). The receptive part of the carpel is called a stigma in the flowers of angiosperms and a micropyle in gymnosperms. The study of pollination brings together many disciplines, such as botany, horticulture, entomology, and ecology. Pollination is important in horticulture because most plant fruits will not develop if the ovules are not fertilised.

Kinds of Pollination

The process of pollination requires pollinators as agents that carry or move the pollen grains from the anther to the receptive part of the carpel. The various flower traits that attract different pollinators are known as pollination syndromes. Methods of pollination, with common pollinators or plants, are:

1. Biotic pollination (by organisms)
   a) Entomophily: pollination by insects
      — Bees, wasps and occasionally ants (Hymenoptera)

- Beetles (Coleoptera)
- Moths and Butterflies (Lepidoptera)
- Flies (Diptera)

b) Zoophily: pollination by vertebrates such as birds or bats

- Hummingbird
- Sunbird
- Honeyeater

2. Abiotic pollination a

a) Anemophily: pollination by wind

- very common in grasses
- Most Conifers
- Many deciduous trees

b) Hydrophily: pollination by water

- Aquatic plants

About 80% of all plant pollination is biotic. Of the 20% of abiotically pollinated species, 98% is by wind and 2% by water.

Pollination Management

Pollination management is a branch of horticulture that seeks to protect and enhance present pollinators and often involves the culture and addition of pollinators in monoculture situations, such as commercial fruit orchards. The largest managed pollination event in the world is in Californian almond orchards, where nearly half (about one million hives) of the US honeybees are trucked to the almond orchards each spring. New York's apple crop requires about 30,000 hives; Maine's blueberry crop uses about 50,000 hives each year.

Bees are also brought to commercial plantings of cucumbers, squash, melons, strawberries, and many other crops. Honeybees are not the only managed pollinators: other species of bees are also raised as pollinators. The alfalfa leafcutter bee is an important pollinator for alfalfa seed in western United States and Canada. Bumblebees are increasingly raised and used extensively for greenhouse tomatoes and other crops.

The ecological and financial importance of natural pollination by insects to agricultural crops, improving their quality and quantity, becomes more and more appreciated and has given rise to new financial opportunities. The vicinity of a forest or wild grasslands near agricultural crops, such as apples, almonds or coffee can improve their yield by about 20%. This may result in forest owners demanding payment for their part in the improved results. This is a simple example of the economic value of ecological services.

The American Institute of Biological Sciences reports that native insect pollination saves the United States agricultural pollination management industry nearly an estimated $3.1 billion annually through natural crop production. Pollination also requires consideration of pollenisers. (The terms "pollinator" and "polleniser" are often confused: a pollinator is the agent that moves the pollen, whether it be wind, bees, bats, moths, or birds; a polleniser is the plant that provides the pollen.)

Some plants are self-fertile or self-compatible and can pollinate themselves. Other plants have chemical or physical barriers to self-pollination and need to be cross-pollinated: with these self-infertile plants, not only pollinators must be considered but pollenisers as well. In pollination management, a good polleniser is a plant that provides compatible, viable and plentiful pollen and blooms at the same time as the plant that is to be pollinated. Pollination can be cross-pollination with a pollinator and an external polleniser, self-pollenisation with a pollinator, or self-pollination without any pollinator:

- *Cross-pollination (syngamy)*: pollen is delivered to a flower of a different plant. Plants adapted to outcross or cross-pollinise have taller stamens than carpels to better spread pollen to other flowers.
- *Self-pollenisation (autogamy)*: pollen moves to the female part of the same flower, or to another flower on the same individual plant. This is sometimes referred to as self-pollination, but this is not synonymous with autogamy. Clarity requires that the term "self-pollination" be restricted to those plants that accomplish pollination without an external pollinator (example: the stamens actually grow into contact with the pistil to transfer the pollen). Most peach varieties are autogamous, but not truly self-pollinated, as it is generally an insect pollinator that moves the pollen from anther to stigma. Plants adapted to self-fertilise have similar stamen and carpel length.
- *Cleistogamy*: pollination that occurs before the flower opens is always self-pollination. Some cleistogamous flowers never open, in contrast to chasmogamous flowers that open and are then pollinated. Cleistogamous flowers must of necessity be self-compatible or self-fertile plants. Other plants are self-incompatible. These are end points on a continuum, not absolute points.

Hybridisation is effective pollination between flowers of different species of the same genus, or even between flowers of different genera (as in the case of several orchids). Peaches are considered self-fertile because a commercial crop can be produced without cross-pollination, though cross-pollination usually gives a better crop. Apples are considered self-incompatible, because a commercial crop must be cross-pollinated. Remember that most fruits are grafted clones, genetically identical. An orchard block of apples of one variety is in effect all one plant. Growers now consider this a mistake. One means of correcting this mistake is to graft a limb of an appropriate polleniser (generally a variety of crabapple) every six trees or so.

Pollination of food crops has become an environmental issue, due to two trends. The trend to monoculture means that greater concentrations of pollinators are needed at bloom time than ever before, yet the area is forage poor or even deadly to bees for the rest of the season. The other trend is the decline of pollinator populations, due to pesticide misuse and overuse, new diseases and parasites of bees, clearcut logging, decline of beekeeping, suburban development, removal of hedges and other habitat from farms, and public paranoia about bees.

Widespread aerial spraying for mosquitoes due to West Nile fears is causing an acceleration of the loss of pollinators. The US solution to the pollinator shortage, so far, has been for commercial beekeepers to become pollination contractors and to migrate. Just as the combine harvesters follow the wheat harvest from Texas to Manitoba, beekeepers follow the bloom from south to north, to provide pollination for many different crops.

Bee Pollination

Bees travel from flower to flower, collecting nectar (later converted to honey), and pollen grains. The bee collects the pollen by rubbing against the anthers. The pollen collects on the hind legs, in dense hairs referred to as a pollen basket. As the bee flies from flower to flower, some of the pollen grains are transferred onto the stigma of other flowers. Nectar provides the energy for bee nutrition; pollen provides the protein.

When bees are rearing large quantities of brood (beekeepers say hives are "building"), bees will deliberately gather pollen to meet the nutritional needs of the brood. A honeybee that is deliberately gathering pollen is up to ten times more efficient as a pollinator than one that is primarily gathering nectar and only unintentionally transferring pollen. Good pollination management seeks to have bees in a "building" state during the bloom period of the crop, thus requiring them to gather pollen, and making them more efficient pollinators. Thus the management techniques of a beekeeper providing pollination services are different from, and somewhat incompatible with, those of a beekeeper who is trying to produce honey.

Hand Pollination

Hand pollination (also called "mechanical pollenation") is a technique used when natural, or open pollination is insufficient or undesirable. The most common techniques are for crops such as cucurbits, which may exhibit poor pollination by fruit abortion, fruit deformity or poor maturation. Hand pollination is only an option on a small scale, but is a common technique by gardeners who transfer pollen with an artist brush or cotton swab. Sometimes the corolla is removed from male flowers and the flower itself is brushed against the stigmas of female flowers.

Hand pollination can be the simple act of shaking, with tomatoes and other self fertile flowers. Tomatoes grown in greenhouses generally need aid in pollination, and this has been done by hand or with an electric vibrator for a long time. Recently it has been mostly supplanted by cultured bumblebees. When pollination is needed on a large scale, such as for field crops, orchards or commercial seed production, honeybees or other pollinators are cultured and provided for pollination management.

The primary reason for hand pollination is lack of pollinators which may be due to crops being moved from home areas (as in vanilla beans) without also taking the natural pollinator. Pollinator decline or the concentrated pollination needs of monoculture may also be factors in pollination shortage. Another reason for hand pollination is in order to control the parents of the seed to be produced. For example, all F1 hybrids are the result of pollenating one specific variety with another specific variety, in order to produce a uniform crop.

For pest prevention with some high-value crops, row covers of mesh are used to prevent pest insects from access to the plants. However, this may necessate hand pollination when bloom starts. Hand pollination is used with date palms to avoid wasting a lot of space and energy growing sufficient male plants for adequate natural pollination. Variable techniques for hand pollination have arisen for many specialty crops.

Self-pollination

Self-pollination is a form of pollination that can occur when a flower has both stamens and a pistil in which the cultivar or species is self fertile and the stamens and the sticky stigma of the pistil contact each other to accomplish pollination. The term is inaccurately used in many cases where an outside pollinator is actually required; such plants are merely self fertile, or self pollenising. Few plants actually self pollinate. The mechanism is seen most often in some legumes such as peanuts. In another legume, Soybeans, the flowers open and remain receptive to insect cross pollination during the day; if this is not accomplished, the flowers self pollinate as they are closing.

Self pollination, or more generally self pollenising, limits the variety of progeny and may depress plant vigor. However, self pollenising can be advantageous, allowing plants to spread beyond the range of suitable pollinators or produce offspring in areas where pollinator populations have been greatly reduced or are naturally variable.

Fruit Tree Pollination

Apple Pollination

Apples are self incompatible and must be cross pollinated. Pollination management is an important component of apple culture. Before planting, it is important to arrange for pollenisers—varieties of apple or crabapple that provide plentiful, viable and compatible pollen. Orchard blocks may alternate rows of compatible varieties, or may plant crabapple trees, or graft on limbs of crabapple. Some varieties produce very little pollen, or the pollen is sterile, so these are not good pollenisers. Good-quality nurseries have polleniser compatibility lists.

Growers with old orchard blocks of single varieties sometimes provide bouquets of crabapple blossoms in drums or pails in the orchard for pollenisers. Home growers with a single tree and no other variety in the neighbourhood can do the same on a smaller scale. During the bloom each season, apple growers usually provide pollinators to carry the pollen.

Honeybee hives are most commonly used, and arrangements may be made with a commercial beekeeper who supplies hives for a fee. Orchard mason bees are also used as supplemental pollinators in commercial orchards. Home growers may find these more acceptable in suburban locations because they do not sting. Some wild bees such as carpenter bees and other solitary bees may help. Bumble bee queens are sometimes present in orchards, but not usually in enough quantity to be significant pollinators.

Symptoms of inadequate pollination are small and mishapen apples, and slowness to ripen. The seeds can be counted to evaluate pollination. Well-pollinated apples are the best quality, and will have seven to ten seeds. Apples with fewer than three seeds will usually not mature and will drop from the trees in the early summer. Inadequate pollination can result from either a lack of pollinators or pollenisers, or from poor pollinating weather at bloom time. It generally requires multiple bee visits to deliver sufficient grains of pollen to accomplish complete pollination.

Citrus Pollination

Many citrus varieties are seedless and are produced parthenocarpically without pollination. Some varieties may be capable of producing fruit either way, having seeds in the segments, if pollinated, and no seeds if not. Citrus that requires pollination may be self compatible, thus pollen must be moved only a short distance from the anther to the stigma by a pollinator. Some citrus, such as Meyer Lemons, are popular container plants. When these bloom indoors, they often suffer from blossom drop because no pollinators have access.

Hand pollination by a human pollinator is a solution, though it is important to learn whether the variety is self fertile or self incompatible. A few citrus varieties, including some tangelos and tangerines are self incompatible, and require cross pollination. Pollenisers must be planned when groves are planted. This last group generally requires the addition of managed honeybee hives at bloom time for adequate pollination.

Pollination Syndrome

Pollination syndromes are suites of traits of flowers aimed at attracting a particular type of pollinator. The traits include flower shape, size, colour, reward type and amount, nectar composition, timing, etc. For example, tubular red flowers with copious nectar attract birds; nasty smelling flowers attract flies, etc. The syndromes are the product of convergent evolution, in response to similar selection pressures.

Abiotic Pollination Syndromes

These don't aim to attract animal pollinators. Nevertheless, they have suites of shared traits.

Wind pollination

Flowers may be small and inconspicuous, green and not showy. They produce enormous amounts of tiny pollen grains (hence wind-pollinated plants may be allergens, but seldom are animal-pollinated plants allergenic). They have large feathery stigmas to catch the pollen grains. They grow in low-diversity stands and are among the taller species in their communities. Insects may visit them to collect pollen, but they are not the most effective pollinators and exert little selection pressure on them.

Water pollination

Water-pollinated plants are aquatic. Their flowers tend to be small and inconspicuous with lots of pollen grains and large, feathery stigmas to catch the pollen. Many aquatic plants are insect-pollinated, with flowers that emerge into the air.

Biotic Pollination Syndromes

Bee pollination

Bee-pollinated flowers tend to fall into two classes:

- Showy, open, bowl-shaped flowers that are relatively unspecialised (e.g. wild roses, sunflowers)
- Showy, complicated, non-radially symmetric flowers that are more specialised (e.g. peas, foxgloves)

Some bee flowers tend to be yellow or blue, often with ultraviolet nectar guides and scent. Nectar, pollen, or both are offered as rewards in varying amounts. The sugar in the nectar tends to be sucrose-dominated. There are diverse types of bees, however. Honeybees, bumblebees, orchid bees, etc are large groups that are quite distinctive in size, tongue length and behaviour (some solitary, some colonial). Thus generalisation about bees is difficult. Some plants can only be pollinated by bees because their anthers release pollen internally, and it must be shaken out by buzz pollination. Bees are the only animals that perform this behaviour.

Butterfly pollination

Butterfly-pollinated flowers tend to be large and showy,

pink or lavender in colour, frequently have a landing area, and are usually scented. Since butterflies do not digest pollen (with one exception), more nectar is offered than pollen. The flowers have simple nectar guides with the nectaries usually hidden in narrow tubes or spurs, reached by the long tongue of the butterfly.

Moth pollination

Among the more important moth pollinators are the hawk moths (Sphingidae). Their behaviour is similar to hummingbirds: they hover in front of flowers with rapid wingbeats. Most are nocturnal or crepuscular. So moth-pollinated flowers tend to be white, night-opening, large and showy with tubular corollas and a strong, sweet scent produced in the evening, night or early morning. A lot of nectar is produced to fuel the high metabolic rates needed to power their flight. Other moths (Noctuids, Geometrids, Pyralids, for example) fly slowly and settle on the flower. They do not require as much nectar as the fast-flying hawk moths, and the flowers tend to be small (though they may be aggregated in heads).

Fly pollination

There are two types of fly pollination: myophily and sapromyophily. A diversity of flies (particularly bee flies (Bombyliidae), hover flies (Syrphidae), etc.) feed on nectar and pollen as adults, and regularly visit flowers. These are the myophiles. Sapromyophiles, on the other hand, normally visit dead animals or dung. They are attracted to flowers that mimic these odoriferous items!

They obtain no reward and would quickly leave, but the plant may have traps to slow them down. These plants have a strong, unpleasant odor, and are brown or orange in colour. They are not as common as myophilous plants. Myophilous plants do not tend to have a strong scent, and tend to be purple, violet, blue, and white, open dishes, or tubes. Flies tend to be important pollinators in high-altitude and high-latitude systems, where they are numerous and other insect groups may be lacking.

Bird pollination

Although hummingbirds are the most familiar nectar-feeding birds for North Americans, there are analogous species in other parts of the world: sunbirds, honeyeaters, flowerpeckers, honeycreepers, bananaquits, flowerpiercers and lorikeets. Hummingbirds are the oldest group, with the greatest degree of specialisation on nectar. Flowers attractive to hummingbirds that can hover in front of the flower tend to be large red or orange tubes with a lot of dilute nectar, secreted during the day. Since birds do not have a strong response to scent, they tend to be odorless. Perching birds need a substantial landing platform, so sunbirds, honeyeaters, and the like are less associated with tubular flowers.

Bat pollination

Bat-pollinated flowers tend to: be large and showy, white or light coloured, open at night and have strong odours. They are often large and bell-shaped. Bats drink the nectar, and these plants typically offer nectar for extended periods of time. Sight, smell, and echo-location are used to initially find the flowers, and excellent spatial memory is used to visit them repeatedly.

In fact bats can identify nectar-producing flowers using echolocation, a talent that was only recently discovered. In the New World, bat pollinated flowers often have sulphur-scented compounds, but this does not carry to other parts of the world. Bat-pollinated plants have bigger pollen than their relatives.

Beetle pollination

Beetle-pollinated flowers are usually large, greenish or off-white in colour and heavily scented. Scents may be spicy, fruity, or similar to decaying organic material. Most beetle-pollinated flowers are flattened or dish shaped, with pollen easily accessible, although they may include traps to keep the beetle longer. Beetles may be particularly important in semi-desert areas, like South Africa and southern California.

Pollination syndromes reflect convergent evolution towards forms (phenotypes) that limit the number of species of pollinators visiting the plant. They increase the specialisation of the plant with regard to pollination. They are responses to common selection pressures exerted by shared pollinators, which generate correlations among traits. That is, if two distantly related plant species are both pollinated by nocturnal moths, for example, their flowers will converge on a form most attractive to the moths (i.e. pale colour, sweet scent, nectar released at the end of a long tube, night-flowering), because the most attractive forms will produce the most offspring.

Advantages

Efficiency of pollination: the rewards given to pollinators, commonly nectar or pollen or both, but sometimes oil, scents, or wax, may be costly to produce. Nectar can be cheap, but pollen is generally expensive as it is relatively high in nitrogen compounds. A plant seeks to obtain the maximum pollen transfer for the minimum reward. Different pollinators, because of their size, shape, or behaviour, have different efficiency of transfer of pollen. And the floral traits affect efficiency of transfer: columbine flowers were experimentally altered and presented to hawkmoths, and flower orientation, shape and colour were found to affect visitation rates or pollen removal.

Pollinator constancy: to efficiently transfer pollen, it is best for the plant if the pollinator focuses on one species of plant, ignoring other species. Otherwise, pollen may be dropped uselessly on the stigmas of other species. Animals, of course, do not aim to pollinate, they aim to collect food as fast as they can. However, many pollinator species exhibit constancy, passing up available flowers to focus on one plant species. Why should animals specialise on a plant species, rather than move to the next flower of any species? Although pollinator constancy was recognised by Aristotle, we don't fully understand the benefits to animals. The most common hypothesis is that pollinators must learn to handle particular types of flowers, and they have limited capacity to learn different types. They can only efficiently gather rewards from one type of flower.

Pollinators fluctuate in abundance and activity independently of their plants, and any one species may fail to pollinate a plant in a particular year, thus a plant may be at an advantage if it attracts several species or types of pollinators, ensuring pollen transfer every year. Plants do, in many species, have the back-up option of self-pollination, if they are not self-incompatible.

9

Genetic Engineering of Crops

The term biotechnology refers to a broad spectrum of technologies-including conventional plant selection and breeding-in which humans intervene in biological processes of genetic alteration and improvement.

The products of genetic engineering are often called genetically modified organisms, or GMOs. All these terms refer to methods of recombinant DNA technology by which biologists splice genes from one or more species into the DNA of crop plants to transfer chosen genetic traits. Genes are segments of DNA that contain information that in part determines the structure and function of a living organism. Genetic engineers are molecular biologists who manipulate this information, typically by taking genes from one species-an animal, plant, bacterium, or virus-and inserting them into another species, such as an agricultural crop.

With the advent of genetic engineering of plants around 1983, it appeared that this new biotechnology would benefit and even revolutionise agriculture. The transfer of desirable genetic traits across species barriers has shown promise for solving problems in the management of agricultural crops. Potential benefits include reduced toxic pesticide use, improved weed control resulting in less tillage and soil erosion, and water conservation. However, "emerging evidence suggests the promised environmental benefits remain small, uncertain or unrealised in the U.S., and some risks are real".

Gene Transfer

Knowing how gene transfer occurs in the laboratory is key to understanding how the desired traits express themselves in the resulting crop plants, and why unintended effects, such as the production of allergens, may also be present or develop with subsequent generations of plants. Plants have elaborate defence mechanisms for dealing with foreign substances, including foreign DNA. To overcome these defence mechanisms, genetic engineers include pathological organisms in the complex DNA packages they create to carry the chosen gene into the host cell.

Generally, there are four parts to a DNA package:

1. Genes for the desired trait-the "payload." An example of such a trait is crop resistance to a given herbicide.
2. Genes for carrying the package into the host plant's DNA. This genetic carrier is called the vector and is usually taken from a bacterium that causes tumors in plants, Agrobacterium tumefaciens. Viruses are also sometimes used as gene carriers. These bacterial or viral vectors infect the new host cells, delivering the engineered gene into the DNA of the host plant.
3. Genes for ensuring that the genetic package will express the desired trait persistently (rather than weakly or not at all). These genes "turn on" the desired trait and are called promoters. They are

usually derived from the cauliflower mosaic virus (CaMV).

4. Genes for helping the biologist find the DNA segment in which the insertion has been successful. These genes are called markers and are resistant to antibiotics (usually neomycin or kanamycin). When the antibiotic is applied to the new host's cells, the cells that survive are the ones carrying the successfully inserted antibiotic-resistant gene-indicating a likelihood that the gene carrying the desired trait has been successfully inserted as well. Marker genes are usually derived from bacteria (E. coli).

An alternative vector method uses a particle gun to shoot tiny metal particles with the DNA package attached into the new host cell. This eliminates the need for a bacterial vector, but adds to the potential for genetic disruption in the host cell.

The current methods of gene transfer are not precise. While scientists can control with relative exactness the "trait gene" that is inserted into a host plant genome, they cannot yet control its location, nor the number of copies that get inserted. Location of genetic material is important because it controls the expression of biological traits just as genes themselves do. Also, inserted DNA frequently contains multiple copies, which are often scrambled.

Multiple copies can "silence" genes, resulting in instability in the introduced genes. Because scientists can control neither the location nor copies of the genetic material they are inserting, they cannot entirely control the traits the host plant will express, nor can they guarantee genetic stability in subsequent generations. This can lead to unpredictable and undesirable effects, examples of which include plant infertility, the production of toxins, and reductions in yield and plant fitness. Another characteristic that can result in unintended effects is called pleiotropy.

Pleiotropy means that one gene controls multiple traits in an organism. In other words, the gene for a desired trait may also control other traits in the crop plant. Genetic engineers have relied on a simpler model of gene expression, assuming one gene controls one particular trait But pleiotropy is common, and the interactions of genes with each other and their environment add further complexity. In short, it is difficult or impossible to predict what the effects of new genetic combinations will be.

The current marker and promoter genes of choice also create new hazards. The antibiotic-resistant marker genes carry the potential to increase the number of bacteria that are resistant to antibiotics. The viral promoter genes could combine with other infecting viruses to create new viruses. The very powerful promoter commonly used, the cauliflower mosaic virus, can cause the inserted DNA package to be expressed out of proportion with the rest of the genetic code.

With the particle gun method especially, this promoter can also jump out of the DNA package and land somewhere else in the host genome, causing disruption. The bacterial and viral vector genes can recombine to form active pathogens once again-either new ones, or old ones with renewed virulence, or with broader host specificity. Each piece of the gene package described above carries with it the potential to disrupt non-target portions of the host plant's DNA, to create instability in the new genetic construct, or to result in unpredictable combinations that can create new substances, viruses, or bacteria.

Human Genome Project

Genetic engineering of crops is based upon the notion that genes alone determine species traits. But this assumption-and the resulting idea of simply transferring a single gene between organisms to achieve a single result-contradicts recent discoveries in genetic science. Inheritance and expression of traits are controlled by complex processes involving complicated genetic material that includes more than simple genes. For example, so-called junk DNA is now thought to play a significant role in determining an organism's traits.

Mapping of the human genome, completed in early 2001, revealed that humans contain about one-third the number of genes scientists expected to find. In fact, the 30,000 human genes they identified comprise only 1 percent of the total human DNA; they don't know what much of the other 99 percent is for. Only 300 genes in the human genome are not also in the mouse.

Commercial Transgenic Crops

While increased yields and improved nutritional value are among the promised benefits of transgenic crops, most genetically engineered crops now planted worldwide are designed either to 1) survive exposure to certain herbicides (called herbicide-tolerant crops) or 2) kill certain insect pests (called pesticidal or insecticidal crops).

Genetically engineered herbicide-tolerant crops have been altered to withstand being sprayed with broad-spectrum herbicides, with the idea that one application will take care of most types of weeds without killing the crop. Insecticidal crops contain genes of the soil bacterium Bacillus thuringiensis (Bt). These Bt genes cause the plants to produce a chemical toxic to the

European corn borer, the cotton bollworm, and other caterpillars. (Caterpillars are the larvae of insects in the Lepidoptera order, which includes moths and butterflies).

Herbicide-tolerant crops accounted for 71 percent of the acreage planted, worldwide, to genetically engineered crops in 1998 and 1999. Pesticidal crops, or a combination of pesticidal and herbicide-tolerant crops, accounted for most of the remaining acreage. More than 100 million acres of the world's farmland were planted with transgenic crops in the year 2000. The United States, Argentina, and Canada are the world's leading producers of genetically engineered crops, with the U.S. first at 71 percent of the world's total acreage.

Transgenic crops are being shipped to or experimented with in many other countries, including China, India, Australia, and South Africa. In 1999, one-quarter of U.S. farmland was planted to transgenic crops. To date, soybeans and corn cover the most transgenic acres. In 2001, more than 60 percent of soybeans planted in the U. S. were estimated to be transgenic-specifically, tolerant to Monsanto's broad-spectrum herbicide, Roundup™ (glyphosate).

Transgenic corn planted in the U.S. offers both insecticidal and herbicide-tolerant traits, though Bt insecticidal corn predominates. Other large-acreage transgenic crops include cotton and canola. Most transgenic cotton is herbicide-tolerant, though some varieties have the Bt trait; transgenic canola is herbicide-tolerant. The first transgenic wheat, planned for commercial introduction in 2003, is Roundup-tolerant.

Other traits engineered into commercial transgenic varieties include disease resistance, high pH tolerance, and several nutritional, taste, texture, and shelf-life characteristics. Other transgenic crops currently on the

U.S. market include tomatoes, potatoes, sunflowers, peanuts, and sweet mini-peppers. More transgenic crops, including rice, are under development for commercial release in the next several years. However, there is evidence that biotechnology firms may be changing course because of deepening consumer skepticism and tighter regulation worldwide, which are increasing the costs and business risks of new transgenic crop introductions.

Transgenic-crop-related Issues

Ecological Issues

Gene flow

Ecological scientists have little doubt that gene flow from transgenic fields into conventional crops and related wild plants will occur. Gene flow from transgenic to conventional crops is of concern to farmers because of its potential to cause herbicide resistance in related conventional crops. For example, in western Canada, three different herbicide-resistant canola varieties have cross-pollinated to create canola plants that are resistant to all three types of herbicide. Through what is called gene-stacking, this new triple resistance has turned volunteer canola into a significant weed problem.

Gene flow from transgenic crops to wild relatives creates a potential for wild plants or weeds to acquire traits that improve their fitness, turning them into "super weeds." For example, if jointed goatgrass-a weedy relative of wheat-acquires the herbicide-tolerant trait of Roundup Ready wheat, it will thrive in crop fields unless applications of other herbicides are made. There is already evidence of such outcrossing from herbicide-resistant wheat to jointed goatgrass. Frank Young and his colleagues at Washington State University found that imidazolone-resistant wheat (not a transgenic variety) outcrossed to goatgrass in one season.

Other traits that wild plants could acquire from transgenic plants that would increase their weediness are insect and virus resistance. Because of their experience with classically bred plants, few scientists doubt that genes will move from crops into the wild: seven of the world's thirteen most important crop weeds have been made weedier by genes acquired from classically bred crops. Because gene flow has the potential to affect farmers' crop and pest management, crop marketability, and liability, more research needs to be done to determine the conditions under which gene flow from transgenic plants is likely to be significant.

Pesticide resistance in insect pests

Bt has been widely used as a microbial spray because it is toxic only to caterpillars. In fact, it is a pest management tool that organic farmers depend on-one of the few insecticides acceptable under organic rules. Unlike the commercial insecticide spray, the Bt engineered into crop plants is reproduced in all, or nearly all, the cells of every plant, not just applied on the plant surface for a temporary toxic effect. As a result, the possibility that transgenic Bt crops will accelerate insect pests' development of resistance to Bt is a serious concern. Pest resistance to Bt would remove this valuable and environmentally benign tool from farmers' and forest managers' pest control toolbox.

Antibiotic resistance

The use of antibiotic-resistant marker genes for the delivery of a gene package into a recipient plant carries the danger of spreading antibiotic-resistant bacteria. The likely result will be human and animal health diseases resistant to treatment with available antibiotics. Research is needed on antibiotic resistance management in transgenic crops. Already the European Commission's new rules governing transgenic crops stipulate phasing out antibiotic-resistant marker genes by the end of 2004.

Effects on beneficial organisms

Evidence is increasing that transgenic crops-either directly or through practices linked to their production-are detrimental to beneficial organisms. New studies are finding that Bt crops exude Bt in concentrations high enough to be toxic to some beneficial soil organisms. In addition, a study by University of Arkansas scientists has shown that "root development, nodulation and nitrogen fixation are impaired" in some varieties of Roundup Ready soybeans.

The reason is that the beneficial rhizobium responsible for nitrogen fixation in soybeans is sensitive to Roundup. It also appears that disruption of beneficial soil organisms can interfere with plant uptake of phosphorus, an essential plant nutrient. Beneficial insects that prey on insect pests can be affected by insecticidal crops in two ways. First, the Bt in transgenic insecticidal crops has been shown in some laboratory studies to be toxic to ladybird beetles, lacewings, and monarch butterflies.

The extent to which these beneficials are affected in the field is a matter of further study. Second, because the insecticidal properties of Bt crops function even in the absence of an economic threshold of pests, Bt crops potentially can reduce pest populations to the point that predator species are negatively affected.

Reduced crop genetic diversity

As fewer and larger firms dominate the rapidly merging seed and biotechnology market, transgenic crops may continue the trend toward simplification of cropping systems by reducing the number and type of crops planted. In addition, seed-saving, which promotes genetic diversity, is restricted for transgenic crops.

Food Safety

Following are some consumer safety concerns about transgenic food:

- possibility of toxins in food
- possibility of new pathogens
- reduced nutritional value
- introduction of human allergens
- transfer of antibiotic resistance to humans
- unexpected immune-system and genetic
- effects from the introduction of novel' compounds

It is in part because of these concerns that consumer demand for organically grown crops continues to increase.

Allergenicity

StarLink corn, a transgenic variety, contains the Cry9C protein, which protects the growing plant from pests. In 1998, federal regulators approved Starlink for livestock feed and ethanol, but banned it in human food because of concerns that the protein could cause allergic reactions. Yet, by the fall of 2000, StarLink had found its way into taco shells shipped around the world.

Many people reported allergic reactions after eating foods later discovered to contain StarLink. However, proving a link to StarLink was difficult because the products containing it were not labelled and people ate it unknowingly. Aventis, the company that developed and patented StarLink, asked the U.S. Environmental Protection Agency (EPA) to approve it for human consumption after the fact in order to avoid the costs of removing it from the world's market. In July 2001, a U.S. science advisory panel recommended to the EPA that it maintain its ban on StarLink corn in human food. The science panel reaffirmed that the Cry9C protein in StarLink corn is likely to be a human allergen.

Farm Management Issues

The most widely planted transgenic crops on the market today can simplify short-term pest management for farmers and ranchers. In the case of herbicide-tolerant crops, ideally farmers can use a single broad-spectrum herbicide for all their crop weeds, though they may need more than one application in a season. By planting insecticidal crops, farmers can eliminate the need to apply pesticides for caterpillar pests like the European corn borer or the cotton bollworm, though they still have to contend with other crop pests.

While these crops offer handy pest control features, they may complicate other areas of farm management. Farmers who are growing both transgenic and conventional varieties of the same crop will need to segregate the two during all production, harvesting, storage, and transportation phases if they are selling into differentiated markets or plan to save their own seed from their conventional crops. To minimise the risk of gene flow from transgenic to adjacent conventional crop fields, federal regulations require buffer strips of conventional varieties around transgenic fields.

Different transgenic crops require different buffer widths. Because the buffer strips must be managed conventionally, producers have to be willing to maintain two different farming systems on their transgenic fields. Crops harvested from the buffer strips must be handled and marketed as though they are transgenic. Planted refuges-where pest species can live outside fields of insecticidal and herbicide-tolerant transgenic crops-are also required to slow the development of weed and insect pest resistance to Bt and broad-spectrum herbicides.

These refuges allow some individuals in the pest population to survive and carry on the traits of pesticide susceptibility. Requirements governing the size of refuges differ according to the type of transgenic crop grown. Farmers growing herbicide-tolerant crops need to be aware that volunteer crop plants the following year will be herbicide-resistant. This herbicide resistance makes no-till or direct-seed systems difficult because volunteers can't be controlled with the same herbicide used on the rest of the crop.

In a no-till system that relies on the same broad-spectrum herbicide that the volunteer plants are resistant to, these plants will contaminate the harvest of a following conventional variety of the same crop, a situation farmers need to avoid for two reasons. First, the contamination means a following conventional crop will have to be sold on the transgenic market. This leads to the second reason. If farmers grow and market a transgenic crop for which they do not have a technology agreement and did not pay royalty fees, they face possible prosecution by the company that owns the transgenic variety.

Many farmers have been charged with "theft" of a company's patented seed as a result of contamination in the field. Farmers growing insecticidal crops need to recognise that insect pressure is difficult to predict and may not warrant the planting of an insecticidal variety every year. In a year when pest pressure is low, the transgenic seed becomes expensive insurance against the threat of insect damage.

Farmers growing transgenic crops need to communicate with their neighbours to avoid contaminating their fields and to ensure that buffers are adequate. In Maine, farmers growing transgenic crops are now required by law to be listed with the state agriculture department, to help identify possible sources of cross-contamination when it occurs. The new law, L.D. 1266, also "requires manufacturers or seed dealers of genetically engineered plants, plant parts or seeds to provide written instructions to all growers on how to plant, grow, and harvest the crops to minimise potential cross-contamination of non-genetically engineered crops or wild plant populations". Farm management issues common to all transgenic crops include yield, cost, price, profitability, management flexibility, sustainability, market acceptance, and liability.

Crop Yield, Costs, and Profitability

Some farmers will get higher yields with a particular transgenic crop variety than with their conventional varieties, and some will get lower yields. The same is true for costs. Some producers will see their overall profitability rise, and some will see it drop. If the results of research studies so far on yield, chemical use, costs, crop prices, and profitability of these crops seem contradictory, it is because farmers and farms are not all alike. Neither are the conventional varieties used for comparison. Some yield, cost, and profitability trends do appear to be emerging from the growing body of research data for transgenic crops, however.

Herbicide-tolerant soybeans appear to suffer what's referred to as a "yield drag." Again, in some areas and on some farms this tendency of Roundup Ready soybean varieties to yield less than their comparable, conventional counterparts varies, but overall, they appear to average yields that are five to ten percent lower per acre. Impaired root development, nodulation, and nitrogen fixation likely account for this yield drag. Drought conditions worsen the effects. The bacterium that facilitates nodulation and nitrogen fixation in the root zone apparently is sensitive to both Roundup and drought.

Yields of herbicide-tolerant cotton are reportedly not significantly different from those of conventional cotton. Herbicide-resistant transgenic canola varieties yield less on average than conventional canola varieties. Transgenic canola costs less than conventional canola to produce, but because of its higher yields conventional canola returns more profit per acre.

Insecticidal Bt corn and cotton generally yield higher "in most years for some regions" according to USDA Economic Research Service data from 1996 to 1998. Bt cotton, especially, outpaces yields of conventional cotton by as much as 9 to 26 percent in some cases, though not at all in others. Yield increases for Bt corn have not been as dramatic. Time will tell whether farmers can expect yield increases or decreases in the long run with these and other transgenic crop varieties.

Reduction of using pesticides

One of the promises of biotechnology is that it will reduce pesticide use and thereby provide environmental benefits and reduce farmers' costs. The herbicide-tolerant and insecticidal varieties are designed specifically to meet these goals. Studies estimate a two to three percent decrease in U.S. pesticide use, but the effects vary widely by crop, region, and year. Increased future pesticide use resulting from the buildup of resistance to heavily used herbicides is a long-term concern.

Pesticide use depends on the crop and its specific traits; weather; severity of pest infestations; farm management; geographic location of the farm; and other variables. As a result, conclusions drawn by various studies analysing pesticide use on transgenic crops remain controversial. According to the Wallace Center report, in a review of the data available up through 2000, crops engineered to contain Bt appear to have decreased the overall use of insecticides slightly, while the use of herbicide-resistant crops has resulted in variable changes in overall herbicide use, with increases in use of some herbicides in some places and decreases in others.

The crop for which studies are showing the largest decrease in pesticide use is Bt cotton, with Bt corn resulting in only small changes. Herbicide-tolerant cotton has also resulted in little change in herbicide use. The data for herbicide-tolerant soybeans seems harder to sort out. If farmers abandon integrated pest management, which utilises a variety of pesticide and cultural control methods, in favour of the simplified control offered by herbicide-resistant and insecticidal transgenic crops, then early findings of reduced pesticide quantities and toxicity may not hold over the long run.

Profitability

Farmers need to consider all the factors that determine profitability. No single factor can tell the whole story.

Transgenic crop seeds tend to be more costly, and farmers have the added expense of a substantial per-acre fee charged by the owners of transgenic varieties. These costs have to be considered along with input cost changes-whether herbicide or insecticide use and costs go down, go up, or stay the same.

Market price is another factor: Prices for some transgenic crops in some markets are lower than prices for comparable conventional crops, though rarely are they higher. Farmers need to watch the markets. Some buyers will pay a premium for a non-transgenic product, though as transgenic seeds find their way into conventional transportation, storage, and processing streams, these premiums may disappear along with confidence that "GMO-free" products are in fact truly free of engineered genes.

Marketing Issues

Buyer acceptance is a significant marketing issue for farmers raising transgenic crops. Farmers need to know before they plant what their particular markets will or won't accept. Since most grain handlers cannot effectively segregate transgenic from non-transgenic crops in the same facility, many companies are channelling transgenic crops into particular warehouses. Farmers need to know which ones and how far away those are. Many foreign markets tend to be more leery of transgenic products than domestic markets, though trade in transgenic livestock feed is more liberal than trade in transgenic human food.

The widespread contamination of corn in the U.S. with the Cry9C Bt transgene (StarLink), which is not approved for human consumption, has resulted in even further resistance on the part of buyers to purchasing transgenic products for human food. According to a report in Britain's The Guardian, "No new transgenic crops have been approved by the European Union (EU) since April 1998, and a defacto moratorium on further approvals has been in place since June 1999". Some countries have banned altogether the production and importation of transgenic crops and food products (Sri Lanka and Brazil), while many others have put in place partial bans or mechanisms to slow their approval for importation or production.

While the European Union officially lifted its two-year moratorium on the introduction of new transgenic crops in February 2001, during the debate over labelling and traceability regulations the moratorium remains in effect. Under the proposed new EU requirements, "all foods and animal feed derived from GMOs have to be labelled and, in the case of processed goods, records have to be kept throughout the production chain allowing the GMO to be traced back to the farm". If approved, the new regulations will hamper the export of U.S. farmers' products to the EU because the U.S. does not require traceability or labelling of transgenic crops.

In July 2001, the European Commission adoped a new directive regulating the release of transgenic organisms, including plants, animals, and microorganisms. In its main provisions, European Directive 2001 / 18 / EC requires the labelling of all products containing GMOs-including processed ingredients-and guarantees traceability from field to plate. Long-term monitoring of transgenic crops following their release is required, as well as molecular documentation of their genetic stability. The directive guarantees the public's access to information about the release and content of GMOs, and the opportunity for comment prior to a proposed release.

The directive also stipulates phasing-out of antibiotic-resistant marker genes by the end of 2004. If the Commission's directive is approved by the member states of the European Union, all GMOs will eventually have to satisfy its requirements before going to market. Compliance with the requirement to document genetic stability may be the most difficult provision to meet, since instability is a common characteristic of current DNA transfer methods. However, approval under the old directive for any transgenic crop already on the market in the European Union will be retained.

While the U.S. does not require mandatory labelling of processed food containing transgenic ingredients, the European Union, Russia, Czech Republic, Japan, South Korea, Taiwan, Australia, New Zealand, and Ecuador do. Because many domestic merchandisers of agricultural commodities do not segregate transgenic from conventional crop varieties, it is impossible for them and the farmers that supply them to serve these food markets. Twenty percent of corn and 8 percent of soybeans produced in the U.S. are exported, and more than 80 percent of these crops is used in animal feed.

Overseas market restrictions on transgenic crops in the feed sector are less stringent than those for food. In contrast, 48 percent of U.S. wheat is exported, with nearly half of this food crop going to countries that are resistant to buying transgenic food, particularly Japan and European nations. Because wheat producers are so dependent on exports, they are approaching the introduction of the first transgenic wheat, expected as soon as 2003, with caution.

The Japanese milling industry has made it clear that it does not want transgenic products. As a result, Monsanto has promised not to introduce its Roundup Ready wheat until Japan gives its approval. At least two states, North Dakota and Montana, have considered legislation that would place a state moratorium on the introduction of transgenic wheat. In addition to national and international policies on the use and importation of transgenic crops, many countries' processors and retailers have set their own corporate policies. Major retail chains in Europe and the U.S. have declared their commitment to avoiding the purchase of transgenic products, both feed and food.

For buyers who don't want transgenic products, the likelihood of contamination of non-transgenic products with engineered genes is troublesome. Since 1997, the European Union has effectively barred U.S. corn imports over the possibility that genetically engineered varieties unapproved in the EU have mixed with sanctioned crops. This has cost American farmers access to a $200 million-a-year market. Organic farmers face even bigger marketing and trade risks, since their buyers can accept no transgenic contamination.

The organic industry does have a system for segregation, but recent tests for transgenic material in organic products demonstrate that it is not immune to contamination from conventional systems. The markets organic farmers have enjoyed, and those that producers of non-GE conventional crops could build upon, will likely disappear if contamination of their products continues to spread. Until the issues of trade in transgenic crops and food products, and their labelling and traceability, are resolved in international trade arenas, the markets for U.S. transgenic crops will remain uncertain.

Changes in what any particular country will or won't accept are occurring all the time, and producers of crops for export are well advised to stay informed of these changes. Segregation of transgenic crops and liability for cross-contamination of conventional and organic crops are issues that obviously overlap with those of marketing and trade.

Liability

The need to separate transgenic crops from conventional and organic ones opens farmers to liability for their product at every step from seed to table. Effective systems for segregation do not exist at present, and will be costly to develop and put into place. Farmers may well end up bearing the added costs of crop segregation, traceability, and labelling. In the meantime, farmers who grow transgenic varieties-and, ironically, those who do not-are liable for transgenic seeds ending up where they aren't wanted: in their own non-transgenic crop fields, in neighbours' fields, in truckloads of grain arriving at the elevator, in processed food products on retail shelves, and in ships headed overseas.

Farmers who choose not to grow transgenic varieties risk finding transgenic plants in their fields anyway, as a result of cross-pollination via wind, insects, and birds, which may bring pollen from transgenic crops planted miles away. Besides pollen, sources of contamination include contaminated seed and seed brought in by passing trucks or wildlife. Those farmers whose conventional or organic crops are contaminated, regardless of the route, risk lawsuits filed against them by the companies that own the proprietary rights to seed the farmer didn't buy.

Likewise, farmers who grow transgenic crops risk being sued by neighbours and buyers whose non-transgenic crops become contaminated. Because contamination by transgenic material has become so prevalent in such a short time, all farmers in areas of transgenic crop production are at risk. Insurance, the most common recourse for minimising potential losses because of liability, is not available to the nation's farmers for this risk because insurance companies do not have enough information to gauge the potential losses.

Most of the farmers who have been accused by transgenic seed companies of illegally growing and harvesting their proprietary transgenic varieties have paid fines to the companies rather than go to court to defend themselves. One Canadian grower of non-transgenic canola, Percy Schmeiser, did go to court against Monsanto, and lost in a recent ruling that requires him to pay Monsanto the approximately US $85,000 value of his crop and $13,000 in punitive damages.

In another example, a Texas organic farm's corn, assumed to be GE-free, was purchased by a processor who made it into organic tortilla chips. Only after the product had been sold and shipped to European retailers was it discovered to be contaminated with transgenic corn. The processor had to recall its product at a cost of over $150,000. The processor chose not to sue the organic farmer, but could have. The retailer, in turn, apparently did not sue the processor. Had it done so, the liability for the retailer's loss could have fallen on both the processor and the farmer.

According to a Boston Globe story, no figures on farmers' losses resulting from liability for transgenic contamination exist, but anecdotal evidence suggests that cases against farmers are becoming "far more common". Until laws or legal precedent clarify the extent of farmer liability, farmers would do well to avoid making assumptions or claims about the purity of their non-transgenic products. Furthermore, producers of transgenic crops need to take all possible precautions against spreading pollen and seed to their own and others' non-transgenic fields and markets.

A full risk assessment and legal clarification of the distribution of liability among farmers, seed companies, grain handlers, processors, and retailers is needed before farmers can rest assured that transgenic crops won't result in lawsuits against them. For farmers who have adopted transgenic technology-and for those who have not-in the words of Brian Leahy, executive director of California Certified Organic Farmers, "This technology does not respect property rights". In essence, farmers who grow transgenic crops on some of their fields, and farmers who grow none, risk bearing tremendous liability. This situation won't change until farmers either gain legal protection or until they all grow transgenic crops on all their fields.

Influence on Public Research

While transgenic crop varieties are generally the property of private corporations, those corporations often contract with public-sector agricultural research institutions for some of their development work. In fact, private investment in agricultural research, including germplasm development, has surpassed public investment in recent years. With this shift in funding sources, the following questions become important: Is the private sector unduly influencing the public research agenda? Are corporations directing public research in less socially valuable directions while research on, for instance, sustainable agriculture goes wanting? Are the outcomes of corporate-funded transgenic research and development by our public institutions equitable across the food and agricultural sectors? Is equity even a consideration of our public institutions when they accept this work?

Because intellectual property rights (patents) apply to living organisms, making them private property, the free flow of scientific information that has historically characterised public agricultural research is being inhibited. What are the implications for the future of agriculture and society of the secrecy that now surrounds so much of what was formerly shared public knowledge?

These and other questions need to be addressed by citizens and their public institutions. These issues are of particular concern to farmers and consumers who would benefit from research into alternative technologies that are less costly (in every way), less risky, and more equitable. Equity requires that the economic benefits and risks of transgenic technology be fairly distributed among technology providers, farmers, merchants, and consumers. For long-term sustainability, farmers need research that focuses on farms as systems, with internal elements whose relationships can be adjusted to achieve farm management goals.

In contrast, transgenic crop research so far has focused on products that complement toxic chemical approaches to control of individual pest species. These are products that can be commercialised by large agribusiness or agri-chemical interests and that farmers must purchase every year. This research orientation only perpetuates the cost-price squeeze that continues to drive so many out of farming.

Industry Concentration and Farmers' Right

The broadening of intellectual property rights in 1980 to cover living organisms, including genes, has resulted in a flurry of mergers and acquisitions in the seed and biotech industries. According to the Wallace Center report, "Relatively few firms control the vast majority of commercial transgenic crop technologies. These firms have strategically developed linkages among the biotechnology, seed, and agri-chemical sectors to capture as much market value as possible. However, these tightly controlled linkages of product sectors raise serious issues of market access, product innovation, and the flow of public benefits from transgenic crops...".

Unlike Plant Variety Protection-which does not allow for the patenting of individual genes, but only of crop varieties-Intellectual Property Rights prohibit farmers from saving their seed and undertaking their own breeding programme, and prohibit plant breeders from using the material to create a new generation of varieties adapted to specific regions or growing conditions. By 2000, agri-chemical giants DuPont and Monsanto together owned 73 percent of the corn seed producers in the U.S.. This kind of corporate control and concentration raises the question of whether there remains enough competition in the seed industry for seed pricing to remain competitive.

As additional concentration occurs, how affordable will seed-all seed, not just transgenic seed-be for farmers? This question takes on added gravity as an increasing number of seed varieties become proprietary. Farmers can't save proprietary seed for planting and so must purchase new seed every year. In addition, farmers choosing transgenic varieties must sign a contract with the owner of the variety and pay a substantial per-acre technology fee, or royalty. The anticipated commercial introduction of transgenic wheat represents a dramatic shift in an industry in which farmers still widely save their own seed.

As non-transgenic varieties become contaminated with transgenic ones, even those farmers who choose to stick with conventional varieties will lose the right to replant their own seed. This loss has already occurred in Canada's canola industry, with Monsanto winning its court case against farmer Percy Schmeiser for replanting his own canola variety that had become contaminated with Monsanto's Roundup Ready canola. The adoption of transgenic crop varieties has brought with it an increasing prevalence of contract production.

While contract production can lead to increased value and reduced risk for growers, farmers are justified in their concern about their loss of control when they sign a contract with a private company. Issues associated with contract production of transgenic crops must be considered within the broader context of a sustainable agriculture to include ownership, control, and social equity.

Transgenic Crops Regulation

Much of the controversy over transgenic crops, both internationally and in the U.S., is in part a result of how the U.S. regulates transgenic crops. The federal government has determined that the commercial products of agricultural biotechnology are "substantially equivalent" to their conventional counterparts and that therefore no new regulatory process or structure is needed for their review and approval. Currently, three federal agencies regulate the release of transgenic food crops in the U.S.: the U.S. Department of Agriculture's Animal and Plant Health Inspection Service (USDA-APHIS), the U.S. Environmental Protection Agency (EPA), and the U.S. Food and Drug Administration (FDA).

USDA-APHIS: The USDA looks at how a transgenic plant behaves in comparison with its unmodified counterpart. Is it as safe to grow? The data it uses are supplied largely by the companies seeking a permit for release of the new crop. Under "fast-track" approval, a process in place since 1997, companies introducing a crop similar to a previously approved version need give only 30 days advance notice prior to releasing it on the market. According to the Wallace Center report, APHIS staff estimate that by 2000, 95 to 98 percent of field tests were taking place under simple notification rules rather than through permitting.

EPA: The EPA regulates the pesticides produced by transgenic crops, such as the Bt in Bt corn and cotton. It does not regulate the transgenic crops themselves. In contrast to its regulation of conventional pesticides, the EPA has set no tolerance limits for the amount of Bt that transgenic corn, cotton, and potatoes may contain.

FDA: The FDA focuses on the human health risks of transgenic crops. However, its rules do not require mandatory pre-market safety testing or mandatory labelling of transgenic foods.

Initially, the U.S. regulatory process for transgenic food crops required product-by-product reviews. Now, however, to simplify and speed up the process, new products can be approved based on the experience gained in reviewing earlier products. According to the Wallace Center report, the implication is that "some crops might be approved, or disapproved, without actual field testing". The regulatory process, in fact, may not answer most questions about the environmental and human health risks of commercial production of these crops.

Central to the policy of substantial equivalence is the assumption that only the end product of transgenic technology is of concern-not the process of genetic modification. Canada has adopted a similar approach. Europe and other U.S. trading partners, however, have taken a more conservative approach. They focus on the process of genetic modification-the source of many of the environmental and human health risks of greatest concern.

How these different approaches play out in reality can be summed up simply: The U.S. and Canada assume a product is safe until it is proven to carry significant risk; the European Union, which follows the "precautionary principle," assumes the same product may carry significant risk until it can be proven safe. The science used by the two approaches is not fundamentally different. The difference is in the level of risk the different societies and political systems are willing to accept.

Principle of precaution

The principle of precaution was developed specifically for issues involving ethics and risk. It is described by Katherine Barrett, project director with the Science and Environmental Health Network, as:... a process for decision-making under conditions of uncertainty. The principle states that when there is reason to believe that our actions will result in significant harm, we should take active measures to prevent such harm, even if cause-and-effect relationships have not been proven conclusively....

It has been invoked in many international laws, treaties and declarations on a range of environmental issues including climate change, marine dumping of pollutants, and general efforts towards sustainability [including the 2000 international Cartagena Biosafety Protocol on the transfer of 'living modified organisms'].... There is growing consensus that the precautionary principle has reached the status of international customary law. The primary elements of the precautionary principle are identified by Barrett as:

1. Avoid harms to the environment that are "irreversible, persistent, bioaccumulative or otherwise serious."
2. "Anticipate and prevent potential harms at the source," rather than relying on reactive measures of mitigation, clean-up, or compensation.
3. Recognise the limits of scientific knowledge and don't expect a full and conclusive understanding of potential consequences before taking precautionary action, especially when the potential consequences are long-term, unconfined, and broad-scale.
4. Shift the burden of proof to the developers of potentially hazardous technologies.
5. Finally, some versions of the precautionary principle include analysis of the costs and benefits of precautionary action, though cost-benefit analysis "is not a sufficiently robust decision-making framework to stand alone".

Sustainable agriculture

In contrast to the ecological approach of sustainable agriculture, "the current generation of transgenic crops follows a pest management model like that employed for chemical pesticides-through interventions that are toxic to pests," the Wallace Center report points out. This "single-tool" approach is likely to fail in the long run because pests will successfully develop resistance that allows them to thrive. Standard plant-breeding methods can potentially solve many of the same problems in agriculture that genetic engineers are working on, though there are areas in which genetic engineering can enhance traditional plant breeding.

Armed with the map of an organism's genetic code, scientists can test which genes are in a plant to select more easily which ones to cross-breed. "Before we knew where the genes were, we were still breeding in the dark," according to Steven Briggs, head of genomics for Syngenta, a Swiss biotechnology giant, as quoted in the New York Times. Framework for gauging a technology's impact on agricultural sustainability. The sustainability of any agricultural technology can be gauged in part by answering a series of questions that emerge from the principles of sustainable agriculture. Farmers and ranchers can ask themselves these questions in the context of their own operations to help determine whether adoption of the technology will move them away from, or toward, increased sustainability.

1. Does the technology increase genetic diversity?
2. Does it maintain a positive balance of pests and predators?
3. Does it protect or enhance soil biota?
4. Does it decrease the quantity or concentration of toxins released into the environment?
5. Does it decrease soil erosion?
6. Does it protect non-target organisms?
7. Does it help protect natural habitats?
8. Does it reduce pest populations and viability?
9. Does it increase farmers' yields? Decrease farmers' costs?
10. Does it increase farmers' market control? Management flexibility? Time?
11. Does it provide benefits to consumers? Will consumers accept it?

12. Does it help citizens globally gain better access to food?
13. Does it protect the public's access to information and improve public trust in agriculture?

10

Hybrid Seed Production

Commercial exploitation of hybrid vigor is one of the most important applications of genetics in agriculture. It has not only contributed to food security, but has also benefited the environment. It is the basis of a billion-dollar agribusiness which has generated significant employment opportunities all over the world. Among the various crop species in which hybrid varieties are used commercially, rice ranks very high. Rice hybrids were first commercialized in the late 1970s in China. During the past decade, Vietnam, India, the Philippines, Bangladesh and the United States have also begun the commercial production of hybrid rice.

About 15 million ha (50%) of rice in China are hybrid varieties, producing 103.5 million mt of paddy annually (an average yield of 6.9 mt/ha). The remaining 50%, planted in inbred high-yielding varieties, produces 81 million mt (5.4 mt/ha). Thus, on average hybrid rices in China yield about 27% (1.5 mt/ha) more than the inbred high-yielding varieties. The commercialization of hybrid rice technology in tropical countries has been facilitated by the research initiated in 1979 at the International Rice Research Institute (IRRI). It was soon recognized that

Chinese rice hybrids were not suited to the tropics, and their grain quality was poor compared to the popular inbred high-yielding varieties grown in tropical countries. Therefore, IRRI researchers concentrated on developing suitable parental lines specially for the tropics, using the CMS system which had been found to be effective in China.

Within a decade, some commercially usable CMS and restorer lines and some elite hybrids were identified and shared with NARS for evaluation and utilization. Some of these hybrids were released for commercial cultivation in the Philippines (IR64616H, IR68284H), India (IR64611H, IR64618H, IR69690H), Vietnam (IR64615H, IR64616H, IR69690H) in the years 1993-1998. Some IRRI CMS lines, IR58025A and IR62829A, were also used by national programs to develop local hybrids in India (e.g. APHR-1, APHR-2, CORH 2, KRH 2, ADTRH 1, etc) and the Philippines. Concurrently, several private seed companies in Asia were also provided with IRRI-bred parental lines.

Currently, about 610,000 ha are covered with hybrid rices in Vietnam (340,000 ha), India (200,000 ha), Bangladesh (50,000 ha), USA (10,000 ha), Philippines (5,000 ha) and Myanmar (5,000 ha). These have given on average about 20-25% (1 - 1.5/ha) higher yields than the inbred HYVs, thereby contributing towards higher on-farm productivity. Currently, hybrid rice technology is considered a viable option to increase rice yields globally.

Since rice is a self-pollinated crop, hybrid seed production must be based on male sterility systems. These may be cytoplasmic (CMS), thermo-sensitive genic (TGMS) or photo-sensitive genic (PGMS). Most of the commercial hybrids used in China and elsewhere are based on a CMS system, although TGMS and PGMS systems are also becoming used commercially (Yuan, pers. comm.). Hybrid seed production technology involving these male sterility systems is complex and labor-intensive. It also requires a sophisticated seed industry infrastructure.

Hybrid rice seed production technology is different from that for inbred rice seed production, and is more complex than the seed production for many other hybrid crops. Currently, the most popular male sterility system is the CMS (popularly known in China as the three-line system). This utilizes three different lines, namely a cytoplasmic male sterile line (A line), a maintainer (B line), and a restorer (R line). Using this system for the commercial production of hybrid seeds involves two major steps. The multiplication of the A line (female parent) must be carried out before the production of hybrid (F_1) seed.

The other system, which is gaining popularity in China and Vietnam, involves a seed parent which is an environmentally sensitive genetic male sterile (EGMS) or S line. This system requires neither a maintainer (B) nor a restorer (R) line. The seed parent can be multiplied just like any conventional inbred when planted in a fertility-inducing environment. When planted in a sterility-inducing environment, it becomes male sterile and may serve as a seed parent in producing F_1 seed of a hybrid. Any line found to form heterotic combinations with a particular S-line can be used as its pollen parent to produce a hybrid. Since this system involve two lines (TGMS or PGMS, and a pollen parent) to produce the hybrid seed, this system is also known as a two-line system. The three- and two-line systems for hybrid rice seed production are shown in Fig. 1.

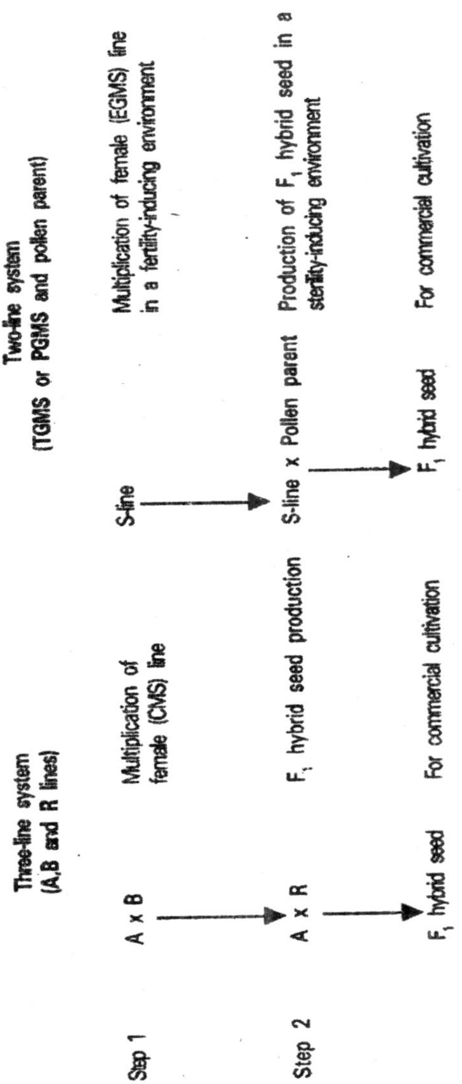

Figure 1. The three- and two-line systems for hybrid rice seed production

For hybrid rice seed production to succeed, a sufficient number of pollen grains must be deposited on the stigma lobes of each spikelet of the seed (male sterile) parent. It helps if the pollen parent grows to a greater height than the seed parent. Other plant characteristics which influence this include small and horizontal flag leaves, the number of panicles per square meter, the number of spikelets per panicle, good panicle exsertion, and synchronized flowering of seed and pollen parents. The floral traits influencing outcrossing in rice include: stigma size, style length, stigma exsertion (in seed parent) and anther length, filament length, and pollen number/ anther (in pollen parent).

The flowering behavior traits influencing outcrossing in rice are: number of days of blooming, time of blooming, duration of blooming, duration of floret opening, and angle of floret opening. The environmental factors influencing outcrossing in rice include temperature, relative humidity, light intensity and wind velocity. The relationship of all these characteristics is discussed at length by Virmani. In China, conditions favorable for good outcrossing in rice have been identified as: a daily temperature of 24-28°C, a relative humidity of 70-80%, a diuranal difference in temperature 8-10°C and sunny days with a breeze. Suitable field conditions include: fertile soil, a dependable irrigation and drainage system and a lcw risk of disease and insect infestations.

Guidelines for Hybrid Rice Seed Production

Seed yield from a male sterile line used in hybrid rice seed production is a function of three factors. First is the yielding ability of the male sterile line, as determined by the yielding ability of its maintainer, the fertile counterpart. Second is the proportion of the male sterile line to the pollen parent. The third is the outcrossing rate of the male sterile line. Improvement in any of these three factors can help to increase the seed yields of hybrid rice.

Extensive research has led to the identification of the following guidelines for successful hybrid rice seed production:

1. Sėlection of seed and pollen parents with synchronized time of anthesis.
2. Selection of seed parents with long, exserted stigma, longer duration, and wider angle of floret opening.
3. Selection of a pollen parent with a high percentage of residual pollen per anther after anther exsertion. High pollen shedding potential is attained by getting 2000 - 3000 spikelets/m^2 to bloom per hour during peak flowering period.
4. Synchronization of flowering time of the two parents by seeding them at different dates depending on their growth duration or estimated accumulated temperature requirements for initiatiion of flowering.
5. Use of optimum seed parent:pollen parent row ratio such that the ratio of spikelet number per unit area of seed parent and pollen parent is about 3.5:1.
6. Use of seed and pollen parents with small and horizontal flag leaves, or cutting long and erect flag leaves.
7. Use of gibberellic acid (GA^3) to improve panicle exsertion and prolong duration of floret opening and stigma receptivity.
8. Planting of seed parent pollen parent rows across the prevailing wind direction and use of

supplementary pollination with a rope or stick when wind velocity is below 2.5 m/sec.

9. Selection of optimum time of flowering of parential lines in seed production plots.

Practices for Hybrid Rice Seed Production

Practices involved in producing hybrid rice seeds using a male sterility system include the following:

Use an Isolated Field

A field should be used which is isolated from other rice varieties by distance (50-100 m), time of flowering (at least 3 weeks) or some natural or artificial barrier. The isolation distance can be decreased by increasing the number of pollen parent rows around the seed production plot. Barrier isolation can be effected through topographical features, or by the use of tall plants or artificial obstacles to block pollen movement.

Timing of Seeding of Parent Lines

The two parental lines should be sown and transplanted at the right times so that their flowering is synchronized. This is the most critical step in hybrid seed production, and is practiced by using one of the three methods described below:

— The seeding date of the two parents can be determined according to the difference in the time they take to reach maturity. The parent which matures earlier must be sown later than the other parent.

— The seeding date of the two parent's can be determined by the difference in the number of leaves at the main culm. If this trait is stable, the date of seeding of the early heading parent can be decided according to the predetermined number of leaves present on the later heading parent.

— The differential seeding of parental lines can be based on their effective accumulated temperature (EAT) from seeding to flowering. The EAT of a particular variety is calculated by summing up the mean temperature of days with a range within 12°C to 27°C. The parent which has a higher EAT is sown earlier, and the other parent is sown after the difference in EAT between the two has accumulated.

Optimum Ratio of Seed Parent

The optimum seed parent:pollen parent ratio must be used, according to the spikelet number per panicle of parental lines, and the size of the pollen load of the pollen parent. The seed parent is planted densely, to increase the number of spikelets per square meter. The pollen parent is more widely spaced, to promote tillering and help ensure that the pollen load is greater over a longer period.

The optimum ratio for plants of the seed: pollen parent should ensure 3.5 spikelets of the seed parent for each spikelet of the pollen parent. Some commonly used seed parent:pollen parent ratios are 10:3, 6:2 and 14:2. In extreme cases, a ratio of 16:2 has been used. In the United States, a ratio of 18:6 is used for mechanized hybrid seed production.

Cultural Management

The seed production plot must be given good cultural management. Weeds, diseases and insects must be controlled, and the optimum fertilizer rate must be applied. If the panicle initiation stage of the two parental lines is found to occur at different times, they should be given different application rates of nitrogen, phosphate and potassium fertilizer, to adjust the flowering date of the parental lines by 2-5 days.

If the parental lines have long, erect flag leaves which might obstruct the movement of pollen, the flag leaves should be clipped by one-half or two-thirds of their length, one to two days before the initial heading. However, this practice has the negative effect that it may spread diseases such as bacterial blight, and sheath blight through the wounds inflicted on the leaves. Besides, this practice also reduces photosynthetic activity. Parental lines possessing short narrow, flag leaves do not need to be clipped.

GA^3 should be sprayed on parental lines which show poor panicle exsertion. Besides improving panicle exsertion, GA^3 also increases the stigma exsertion rate, as well as the time during which the spikelet is open and the stigma is receptive. Differential application of GA^3 also increases the relative height of the pollen parent over the seed parent. GA^3 is generally sprayed using a knapsack sprayer. Spraying begins when 5-10% of the rice plants are at the heading stage, and lasts for 2-4 consecutive days. Spraying should be done on a calm sunny day, usually in the afternoon. It should not be carried out if rain is expected within 24 hours.

Hybrid rice seed producers in China use a very high dosage (150-300 gm/ha) to get high seed yields. However, outside China the high cost of GA^3 limits seed growers to using only 45-50 gm/ha. The use of an Ultra Low Volume (ULV) sprayer has been found to reduce GA^3 dosage to 15-20 gm/ha, and give seed yields similar to those from spraying 45-50 gm GA^3 from a knapsack sprayer.

The horizontal dispersal of pollen grains can be increased by shaking the panicles of the pollen parent by pulling a rope between two people, or stirring the canopy layer of the pollen parent with a bamboo stick (supplementary pollination). This should be done during the anthesis of the pollen parent. Either method may increase outcrossing, and hence seed yields. Care should be taken that panicles are not broken at the neck.

Roguing of off-types should be carried out at maximum tillering, flowering and maturity stages, to prevent any cross-pollination between them and the true-to-type seed and pollen parent plants.

Pollen and seed parents should be harvested, threshed, dried and cleaned separately and carefully, to avoid any mechanical mixture of seed of parental lines.

Using the above practices, seed yields of 0.5 to 4 mt/ha (mean 1.3 mt/ha) have been obtained in the tropics. However, because of the vast experience of China's seed growers and specialized hybrid rice seed production, yields in China range from 1.5 mt/ha to 7.4 mt/ha (mean 2.5 mt/ha). In the late 1970s, average hybrid rice seed yield in Mainland China were 0.67 mt/ha. Currently, these have increased to 2.5 mt/ha. This increase in seed yield is a associated with the modified practices shown in Table 1.

Quality Control

Quality control is vital to the widespread adoption of hybrid rice. Farmers who buy expensive hybrid seeds expect high quality. Quality control is carried out throughout the whole process of seed production. It includes the nucleus, breeder, foundation and certified seed production of both parental lines and F_1 hybrids.

Table 1. Progressive modification of techniques of hybrid rice production in China during the late 1970 to late 1990s to increase seed yields

Technique	Period		
	Late 1970s	Late 1980s	Late 1990s
Row ratio	1-2:6-8	2:10-12	2:14-18
Width of a female parent strip (m)	1.5	2.0	2.5
Plants/hill			
A line	1	2	2
R line	1	1	3-4
Transplanting density (cm)			
A line	16.5 x 16.5	15 x 15	13 x 13
R line	20 x 20	20 x 20	17 x 30
Hills/ha (x 1000)			
A line	320	450	590
R line	67	50	27
Plants/ha (x 1000)			
A line	320	900	1060
R line	67	50	100
Plants/ha (x 1000)	2100	2600	3200
Dosage of GA_3 (g/ha)	0-45	75-120	180-220
Concentration (ppm)	20-40	40-60	100
Seed set percentage	10-15	25-30	45-50
Clipping of flag leaves	Yes	Yes or no	No or lightly
Yield (mt/ha)	0.54	1.80	2.50

Seed quality standards are met by intensive roguing of off-types (from seeding to harvesting) and careful handling during harvesting, threshing, drying, cleaning, processing, bagging and labeling. Grow-out tests, germination tests, seed vigor tests etc. should also be done before the seeds are sold commercially.

China has a well-established hybrid rice seed industry and national standards for hybrid rice. Seed quality standards are in place (Table 2) and being followed. In other countries where hybrid seed production is more recent, standards vary and are still to be formalized.

Vegetative Propagation

Vegetative propagation of rice is well-known. Adult plants can be raised from seedlings, tillers, culm cuttings or by ratooning. Clonal propagation of rice was proposed in early sixties to exploit hybrid vigor in rice. However, no serious attempt was made to develop this as a technology. Since the commercialization of hybrid rice technology in China in 1976, agronomists and farmers have tried using vegetative propagation in various ways to reduce the seed rate, and hence the seed cost, of commercial rice hybrids. These approaches include double transplanting and ratooning.

In double transplanting, hybrid seeds are sown densely in the seed bed. When the seedlings reach the 2-3 leaf stage, they are transplanted into a field nursery, using a wide spacing of 33 x 10 cm to induce more tillering. After 25-30 days, tillers are separated and transplanted into the main rice field. Although double transplanting can reduce the seed rate of hybrid rice significantly, it does increase the labor cost.

In subsistence farming, where family labor is readily available and there is no other opportunity cost, this

Table 2. Seed standards for parental lines and F_1s of hybrid rice in China

	Seed grade	Purity (>%)	Cleanness (>%)	Germination (>%)	Moisture (>%)	Weed seeds (> grains/kg)
A line	Nuclear seed	100.0	99.8	98.0	13.0	0
	Foundation seed	99.9	99.0	96.0	13.0	0
	1st class seed	99.5	99.0	96.0	13.0	0
	2nd class seed	99.0	97.0	95.0	13.0	5
B line	Nuclear seed	100.0	99.8	98.0	13.0	0
	Foundation seed	99.9	99.0	96.0	13.0	0
	1st class seed	99.5	99.0	96.0	13.0	0
	2nd class seed	99.0	97.0	93.0	13.0	5
R line	Nuclear seed	100.0	99.8	98.0	13.0	0
	Foundation seed	99.8	99.0	96.0	13.0	0
	1st class seed	99.5	99.0	96.0	13.0	0
	2nd class seed	99.0	97.0	93.0	13.0	5
F_1 hybrid	1st class seed	98.0	98.0	93.0	13.0	0
	2nd class seed	96.0	97.0	90.0	13.0	5

Note: 1. The 1st and 2nd class seeds are certified seeds.
2. The sterility and sterile plants of nuclear and foundation seeds of A line both should be 100%.
3. The restoring rate of nuclear and foundation seeds of R line both should be higher than 85%.

method may be useful. However, in situations where labor is either not available or can earn more elsewhere, this approach may not be economical.

Ratooning of hybrid rice has been tried with varying degrees of success. Recently, successful vegetative propagatioin of rice has been achieved by splitting and transplanting the ratoon seedlings of an F1 rice hybrid in Bangladesh (M.A. Mazid, pers. comm..). This approach would be practical with resource poor farmers who can use family labor but who cannot afford the full cost of hybrid seed.

Economic Viability

The availability of quality seed at a reasonable price is a crucial factor behind the large-scale adoption of hybrid technology for any crop. This has been demonstrated by the success of hybrid varieties of maize, pearl millet, sorghum, sunflower and cotton, all of which are supported by economical and efficient seed production and distribution systems in the countries where the hybrids are commercialized. Chinese success in the large-scale adoption of hybrid rice can also be attributed to the efficient and economical hybrid seed production and distribution system, organized by the state-owned seed industry.

Hybrid rice seed production technology is considered labor- and knowledge-intensive. It involves various risks, especially in the early stages when seed producers are still lacking in experience. Typical problems are poor synchronization of the parental lines, and unfavorable weather. At present, commercial production of hybrid rice seed outside China is limited to India, Vietnam, the United States and the Philippines. Seed growers will engage in hybrid rice seed production only if it is more profitable than inbred varieties, and if the additional profit is enough to compensate for the greater risks and level of skill involved.

Studies conducted in the Indian states of Andhra Pradesh, Karnataka and West Bengal showed that on average, labor accounted for about 48% of the total cost of hybrid rice seed. Hybrid rice seed production requires about 340 mandays/ha, about 50 mandays/ha more than for ordinary rice cultivation. Hybrid seed production requires additional labor for extra farm operations such as thin and row planting, supplementary pollination, filling gaps, roguing, GA^3 application, and manual harvesting, threshing and cleaning.

In China, the production of hybrid rice seed takes about 375 mandays/ha, 75 mandays/ha more than ordinary rice. The cost of female parental line seed (A seed) and gibberellic acid (GA^3) (additional inputs for hybrid seed production) accounted for 9% of the total cost (Table 3). The average cost of producing hybrid seed was US$0.52/kg. The private sector was able to buy hybrid seed from seed growers at US$0.85/kg, while government agencies paid US$1.0-1.1/kg. At an average price of US$1.0/kg, with an average seed yield of 1250 kg/ha, hybrid rice seed production was 65% more profitable than ordinary rice cultivation (Table 3).

Seed companies in India sell hybrid seed produced in India to farmers at US$2.0-2.5/kg. This large margin is needed to keep the private seed companies in business, since the demand for hybrid rice seed is small and the market is dispersed. However, as the technology picks up, economies of scale should reduce the cost of production and increase seed yields. This would allow seed companies to sell hybrid seed at a lower price to farmers. Greater competition in the seed business will

Table 3. Comparative cost-return profile for hybrid rice seed production and commercial inbred rice production in India (1997-98 crop year)

Costend returns	Hybrid seed production (US$/ha)	Inbred rice production (US$/ha)
Cost		
Seed[a]	70.0	14.0
Fertilizers & manure	123.8	86.8
Gibberellic acid (GA_3)	50.0	-
Labor	315.0	112.5
Plant protection	15.0	13.6
Miscellaneous	71.2	11.9
Total input cost return	645.1	239.0
Yield[b] (kg/ha)	1250	5960
Market price[c] (US/kg)	1.0	0.1173
Return to hybrid seed	1250	-
Return to by-product (male seed and straw)	212.5	41.1
Gross return	1462.5	739.0
Net return	817.4	500.0
Unit cost of production (US$/kg)	0.516	0.040

a: A-line seed in the case of hybrid seed production b: Hybrid seed yield in the case of hybrid seed production
c: Seed price paid to the growers of hybrid seed

also force seed companies to reduce the selling price of hybrid rice. As reported in Table 4, the selling price of hybrid rice seed could be reduced from US$2.50 to 1.0/kg, if seed yields were to increase from their present level of 1.25 mt/ha to 3.0 mt/ha. In such a case, the cost of hybrid seed would fall by US$20-30/ha.

Vietnam, because of its socio-economic and political structure, has been able to develop a strong research-extension-farmer linkage in seed production and distribution, similar to that of China. The top priority for Vietnam's hybrid rice program is to develop an efficient system for the production and distribution of hybrid seed at affordable prices. At present, about 20% of hybrid rice seed requirements are met by seed produced in Vietnam, while the remaining 80% is imported from China. About 600 ha in Vietnam was under hybrid rice seed production in the year 2000, compared to only 173 ha in 1992. Average seed yields increased from about 300 kg/ha to 2500 kg/ha during this period. According to Vietnam's Ministry of Agriculture and Rural Development, hybrid rice seed requirements are about 24,000 mt, if hybrid rice cultivation is to be increased to 800,000 ha by 2005.

The total cost of hybrid rice seed production in Vietnam for the spring crop in 2000 was about US$1200/ha, of which the government provided 30% as a subsidy in the form of free parental line seeds, GA^3, chemical fertilizers, etc. Therefore, the cash cost of seed production paid by the growers was US$800/ha (Table 5). Average seed yields are 2.0 mt/ha, and the current seed procurement price is US$0.60/kg. Thus, the production of hybrid rice seed in Vietnam generates about US$400/ha net profit.

This is 30-35% higher than the profits from ordinary rice. However, the govern-ment's subsidy for hybrid rice

Table 4. Effects of increase in seed yield of hybrid rice on cost of seed production and selling price of seed in India, a sensitivity analysis

Scenario	Cost of production (US$/kg)	Selling price (US$/kg)	Seed cost for commercial cultivation @ 20 kg/ha (US$/ha)
I. At current seed yield of 1,250 kg/ha	0.52	2.50	50
II. At seed yield of 2,000 kg/ha	0.32	1.60	32
III. At seed yield of 2,500 kg/ha	0.26	1.30	26
IV. At seed yield of 3,000 kg/ha	0.22	1.00	20

Table 5. Cost-return profile for hybrid rice seed production in Vietnam, 2000, spring

Particulars	Values
Yield of hybrid seed (kg/ha)	2000
Seed price paid to the seed growers (US$/kg)	0.60
Return to hybrid seed growers (US$/ha)	1200
Total cash costs (US$/ha)	800
Net return to hybrid seed growers (US$/ha)	400
Cost of seed production **with** subsidy (US$/kg)	0.40
Cost of seed production **without** subsidy (US$/kg)	0.60

seed production may not continue forever. If the subsidy is withdrawn, the procurement price of hybrid seed ought to be raised from its present level of US$0.6/kg to 0.8/kg, to compensate. This may lead to an increase of 30% in the selling price, over the current selling price of $1.1-1.2/kg. In order to keep the selling price at the current level without subsidy, the seed yield would have to increase to 3 mt/ha.

The production of hybrid rice seed must be economically viable if it is to attract seed companies. Therefore, seed production can be organized in countries where labor is available at a reasonable price, provided there is a demand for hybrid seed among farmers.

Future Impacts

Successful development and large-scale dissemination of hybrid rice technology will have a major impact on the seed industry, in view of the fact that rice is a staple food crop in Asian countries. Usually, the private sector does not play much of a role in the early stage of technology development. At this stage, the public sector has to play a leading role.

Once the private sector is convinced that there is a demand for hybrid seed, it is willing to invest in commercial seed production, since hybrid seed production is quite profitable. Public sector research institutions in Asia have already made great progress in the development of suitable rice hybrids, and have also developed many parental lines that are freely available to the private sector. A number of private seed companies (including 15 in India, six in Bangladesh, three in the Philippines, and one in Vietnam), and a few NGOs and farmers' organizations, have already begun to produce hybrid rice seed.

There is a growing linkage between the public and private sectors, not only in sharing genetic resources, but also in the form of private sector support for public research. Some small seed companies have also started joint ventures with big multinational com-panies for the production and/or import and marketing of hybrid rice seed in Bangladesh, Vietnam, the Philippines etc..

Some farmers' organizations have also begun hybrid rice seed production, either independently or as a joint venture with a private seed company. These developments may reduce the cost of hybrid rice seeds to as little as US$20-35/ha. This would promote the further adoption of hybrid rice production by small-scale farmers.

In addition to its profitability, the additional labor cost of hybrid rice seed production is a positive feature in rural Asia. Hybrid rice seed production requires about 50 mandays/ha more labor than normal non-mechanized rice cultivation. It therefore generates additional employment opportunities. This is a great benefit, especially for landless women in countries such as India, Bangladesh, and Vietnam which have a labor surplus.

If hybrid rice technology spreads on a large scale in these countries, it would contribute substantially not only to food security, but also to poverty elimination by generating employment in rural areas. If six million hectares were to be planted in hybrid rice by the year 2010, hybrid rice seed production would generate about 3-4 million mandays of rural employment in Asia.

Hybrid rice seed production is quite complex. Many factors affect the yield and quality of hybrid seed. When China began its commercial production of hybrid seed in the late 1970s, the average yield was less than 0.5 mt/ha. It took many years of experience to achieve its current seed yield of 2.5 mt/ha of F_1 seed production in an area of about 0.14 million ha. Twenty years after China, several other countries began the commercial hybrid production of hybrid rice seed. In a few small areas, seed yields of 4 mt/ha have been achieved, but average yields still remain low at around 1.2 mt/ha. There are many constraints to the adoption of hybrid rice seed production technology. These are follows:

1. Hybrid rice technology in countries outside China was adopted only a few years ago. Seed growers need time to become familiar with the technology, and need more experience of hybrid seed production at specific locations under local climatic conditions.
2. There is a shortage of skilled manpower at the initial stages of seed production technology. Seed production can only be successful if it is supervised by well-trained technicians and carried out by experienced growers.
3. Hybrid rice seed production is highly dependent on use of gibberellic acid (GA^3). Outside China this is quite expensive (more than US$1.00 per gram), because it is imported. In China it is quite cheap (US$0.30 per gram) because it is produced locally, so it does not cost much to apply a high dosage of this growth regulator.
4. The purification of parental lines is vital to the success of any hybrid seed production program. However, it has not received enough attention. Consequently, parental lines used for hybrid seed production are not pure, which results in poor-quality hybrid seeds.
5. Hybrid rice seed production requires a well-organized seed industry. Apart from China, most

countries in Asia have not yet achieved this for hybrid rice seed.

6. Since hybrid rice technology is new in most countries, government policies do not yet motivate and encourage hybrid rice seed production on a large scale.
7. The linkage between public sector research institutes and seed production agencies working on hybrid rice is weak.

11

Cultivar Development

Wheat belongs to the Gramineae family. This family includes cultivated wheats, ryes, barleys, oats, and important forage grass species. Within the Gramineae family, the genus Triticum includes a number of species that form a series based on multiples of seven related chromosomes (diploids = 7, tetraploids = 14, hexaploids = 21). Common wheat (*Triticum aestivum*) is a hexaploid (ABD) derived from three diploid species: *Triticum monococcum* (A), a modified diploid (B), and *Triticum tauschii* (D).

Durum wheat (*Triticum turgidum* var. durum) is a tetraploid (AB) derived from the two diploid species that it has in common with hexaploid wheat. The diploid Triticum species presumably also have a common seven chromosome ancestor. The close relationships among Triticum species allows for the exchange of genetic material with the result the wheat relatives have provided plant breeders with a valuable source of exploitable genetic variability, especially for disease and insect resistance.

The need for a vernalisation period of several weeks growth at temperatures approaching freezing before normal heading will occur in winter wheat is the major difference between spring and winter wheat. The vernalisation requirement can be controlled by a single gene. Consequently, there are essentially no restrictions on the plant breeder's ability to move genes between spring and winter wheat classes within species.

Wheat was domesticated from its wild relatives nearly 10,000 years ago in what is now the Mesopotamian region of the Near East. Man has continued to select for adapted varieties of wheat since these early attempts at farming. The Swiss of the Neolithic Period cultivated varieties of wheat and, in ancient Greece, kinds of wheat that had been selected for productivity and value as food were described by one of Plato's pupils in writings completed about 300 B.C. The development and application of plant breeding methods has allowed man to further exploit the evolutionary variability that nature provided in wheat.

Prior to 1900, plants with desirable characters were being selected, intermated, and their superior progeny selected for cultivation. The understanding of genetics that arose out of the work set forth by Mendel in 1865 provided for more rapid, controlled manipulation of desirable genetic traits. With this knowledge, modern plant breeding was born at the start of the 20th century. The basic principles of wheat breeding now include:

1) the search for and identification of plants or varieties possessing desired traits such as rust resistance,
2) the development of suitable practical methods of reliably measuring these traits such as greenhouse screening for rust, and
3) the exploitation of the methods of transfer of superior genes for these traits into adapted cultivars like Norstar, which have the remaining necessary favourable economic traits.

Crimean varieties introduced from eastern Europe into Kansas by Russian Mennonites in 1872 and by the United States Department of Agriculture scientists in 1900 provided the basic germplasm for successful production of hard red winter wheat in the Great Plains of North America. Although most of the early introductions and selections are no longer of commercial importance, most of the strains of hard red winter wheat grown on the Great Plains prior to 1969 were developed from hybrids involving selected Crimean cultivars. It is noteworthy that this same material still ranks with the most winter hardy cultivars grown on the Canadian prairies today.

The early winter wheat introductions into the Great Plains region were landrace mixtures rather than pure line cultivars. Subsequent reselection yielded a number of cultivars, including Kharkov 22MC that was released from Macdonald College in Quebec in 1912. Kharkov 22MC, Jones Fife, and Yogo were the main cultivars grown in southern Alberta in the first half of this century (Table 1). These cultivars all had poor milling and baking quality, especially compared to the spring wheat cultivars grown at that time.

In 1949, winter wheat development in western Canada moved beyond the testing and selection of introductions when a winter wheat breeding program was initiated at Lethbridge. In 1961 this program released Winalta, which had superior breadmaking quality but poorer winter hardiness than Kharkov 22MC. The Lethbridge program also released Sundance in 1971 and Norstar in 1977. Norstar represented the successful combination of good yield potential and baking quality of an adapted parent (Winalta) with superior winterhardiness of an introduced parent (Alabaskaja) and it soon replaced all other winter wheat cultivars on the Canadian prairies.

Cultivar Development

Plant breeding involves the management of a cultivar development program that integrates the knowledge and priorities of a large number of disciplines so that lines with the potential to be superior cultivars can be identified from among thousands of progeny originating from the intermating of selected parents. Plant breeding programs have long term objectives. They must be organised with the understanding that cultivars originating from crosses among parents selected today will not enter the commercial market for 10 to 15 years.

To be successful, wheat breeders should be part of a team made up of pathologists, cereal chemists, plant physiologists, agronomists, biotechnologists, market analysts, etc., and supporting technicians. However, in today's administrative climate of short term objectives (up to 5 years) and contract research, this "team" concept has little basis in reality. In Canada, many of today's plant breeders have been left with the dilemma of priorising the purchases of expertise from support disciplines (where they continue to exist) using funds from budgets that are rapidly ceasing to exist. Following are two important steps in cultivar development:

1. Selection of superior strains
2. Cooperative testing and registration

Superior strains selection

The purpose of a breeding program is to generate genetic variability from which desirable individuals can be selected to produce improved cultivars. Most breeding programs are based on the premise that large populations must be assessed to provide an acceptable probability that individuals with the desired combinations of parental characteristics will exist in the population. Consequently, as many individuals and lines as resources will permit pass through programs each year.

However, little is gained by carrying inferior types from year to year, so they are discarded as soon as effective selection pressure can be applied for important characters. Selection of parents is the initial step in a breeding program. This is also the most important step because it establishes limits on the genetic variability among the progeny available for selection in subsequent generations. Normally one of the parents in a cross will be an adapted cultivar, such as Norstar, while the other parent will possess an important trait that the adapted cultivar is missing, such as rust resistance.

Winter hardiness, drought tolerance, disease resistance, insect tolerance, straw strength, plant height, resistance to shattering, harvestability, seed size, seed shape, seed colour, test weight, and grain quality are among the many traits that winter wheat breeders must consider when selecting parents in a breeding program. Plant breeders make as many as 100 new crosses each year. Consequently, considerable effort is spent on the evaluation of potential parents.

A high winter survival requirement places a large restriction on a breeder's ability to objectively evaluate most potential winter wheat parents under western Canadian conditions, especially outside of southern Alberta. Because of this restriction, nonhardy parental lines used in western Canadian winter wheat breeding programs are often selected without the benefit of local performance data. In these instances the breeder will rely on performance data from trials grown in regions with milder winter climates. A cross between two parents produces hybrid (F_1) seed.

Normally each F_1 seed is the result of a cross that is made by hand. This is a labour intensive process and, because the F_1 generation is genetically uniform, fewer than 50 F_1 seeds are usually produced for each cross. Biological male sterility and fertility restoration systems and chemical gameticides permit the production of larger amounts of F_1 seed. If the yield potential of the F_1 produced by these methods is high enough to offset increased seed costs and the need to purchase seed each year, then F_1 seed may be sold as hybrid seed.

Hybrid winter wheat cultivars have been developed for commercial production on the USA Great Plains. However, the production of hybrid winter wheat has not yet been given serious consideration in western Canada. In contrast to the genetic uniformity of the F_1 generation, the genetic differences between the parents are expressed in a multitude of combinations that appear in the next (F_2) generation. It is at this stage that the plant breeder may resume the selection process. However, the progeny from selected individuals does not breed true in early generations (F_2 to F_5) and the plant breeder normally reselects within each generation.

The progeny of selected individuals becomes genetically more uniform with each succeeding generation and selection for more complex traits, such as yield, normally begins by the F_6. Increased genetic uniformity and the availability of performance data from small plot trials, as opposed to individual plants or single rows, permits the rapid culling of selected strains as they pass from the F_6 to the F_8 generation. The breeding procedures that have been outlined to this point represent a simplified version of most programs. Depending on the circumstances, wheat breeders may superimpose any of a number of options on this main theme.

Spring crop breeders normally have approximately eight months between harvest and the start of seeding in the spring for data analyses, quality evaluation, decision making, etc. In contrast, winter wheat is harvested in August and seeded before mid-September in most of western Canada. Unless they wait and advance their breeding material one generation every second year, winter wheat breeders have one to six weeks to complete the tasks that spring crop breeders perform in eight months. This time factor means that different approaches must be used in spring and winter wheat breeding programs.

Spring crop breeders may grow more than one generation each year by making use of winter increases in the southern U.S.A., Mexico, New Zealand, or other regions with production seasons that are opposite to that of western Canada. Increases in regions with shorter winters would allow for the seeding of winter wheat as late as November while advancing harvest to June. This change would increase the efficiency of Canadian winter wheat programs by allowing more time for processing of breeding material between harvest and seeding.

However, even with increases in southern U.S.A., winter wheat breeders would still only be able to advance their programs one generation each year. While they are more expensive to use, greenhouses and growth chambers provide the only real option for accelerating a winter wheat breeding program, especially in early generations when population sizes are small.

Cooperative testing and registration

Superior F_8 or later generation strains are advanced to cooperative, or pre- registration trials where they undergo final evaluation for a minimum of three years at 10 to 20 locations scattered throughout western Canada. The Prairie Registration Recommending Committee for Grain (PRRCG) is responsible for coordination of cooperative trials established to evaluate potential winter wheat cultivars for the Wheat Board area of western Canada (Figure 1).

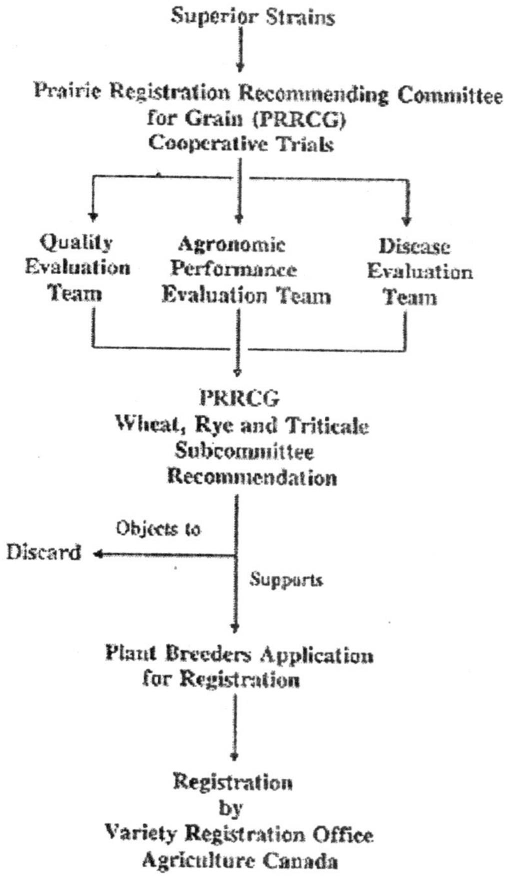

Figure 1. Cooperative testing system for the evaluation of potential cultivars

The data from cooperative trials is used by PRRCG evaluation teams to determine the merit of strains that plant breeders propose for registration. Once a line under evaluation receives the necessary support from the PRRCG, the plant breeder applies to the Variety Registration Office of Agriculture Canada for registration. When a superior winter wheat strain has been identified, the plant breeder will begin a strain purification process to ensure that acceptable breeder seed stocks are available for distribution to seed growers should the strain be successfully registered as a cultivar. This process usually starts no later than after the first year of evaluation in cooperative trials. Breeders seed is distributed to seed growers once a cultivar registration is approved. It then takes at least two more years of increase before seed of the cultivar is available in sufficient quantities for commercial production.

Winter Hardiness

Winter wheat overwinters as a seedling. During the overwintering period the seedling must cope with extremes in winter weather. Many factors can influence winter survival. However, under western Canadian conditions, the main cause of winterkill is simply low temperature damage to critical parts of the plant during periods of cold.

Considerable genetic variability for winter hardiness exists in the winter wheat gene pool. It is this variability that breeders must exploit in efforts to maintain and improve the winter hardiness of winter wheat cultivars developed for western Canada. Hardy winter wheat cultivars are a reality, but whether or not the existing genetic variability can be re-worked to produce super-hardy winter wheats is another question.

In 1929, the American researchers Quisenberry and Clarke observed that "the possibility of developing hardier varieties has been recognised for years." While this general optimism still prevails today, the record shows that since 1929, the total world breeding effort has achieved little or no increase in the winter hardiness of available cultivars (Table 1). Early wheat breeders/ farmers, using trial and error methods, were very effective in selecting the most winter hardy strains, and new major advances have proven difficult to achieve.

Minhardi and Kharkov 22MC were released for commercial production in 1902 and 1912, respectively. They were both reselections from Crimean introductions that arrived in North America in the late 1800's. These two cultivars still rank with the hardiest cultivars available today.

Table 1. Winter hardiness of selected winter cereal cultivars and strains.

Cultivar	Year commercially available in North America	FSI
Jones Fife	1889	401
AC ReadyMade	1991	437
Winalta	1961	463
Triticale (Puma plus Norstar)	N/A	476
Norwin	1986	476
Minhardi	1902	492
Yogo 1932	493	
Sundance	1971	494
CDC Kestrel	1991	497
Kharkov 22MC	1912	499
Norstar	1977	514
Alabaskja	1912	527
Ulianovkia	N/A	530
Puma Rye		735

Alabaskaja and Ulianovkia, two more recent introductions from Russia, have winter hardiness levels superior to Minhardi and Kharkov 22MC (Table 1). Part of this winter hardiness advantage has been captured in the cultivar Norstar, which originated from a cross between Alabaskaja and Winalta. However, both Alabaskaja and Ulianovkia have been available in Russia for many years and, while they have provided North American wheat breeders with additional genetic variability, they do not represent new sources of major genes for winter hardiness.

Related species provide another potential source of genetic variability that may be exploited to improve the winter hardiness of winter wheat. The superior winter hardiness of rye provides proof that the potential to survive winter stresses is much greater than is presently available in wheat (Table 1). However, transfer of this potential from rye to wheat has proven to be a difficult task. For example, the winter hardiness advantage of rye is completely suppressed when winter wheat and rye are combined to produce triticales (Table 1).

Attempts at interspecific transfers of winter hardiness genes to wheat from its close relatives have also been unsuccessful. Advances in biotechnology provide an opportunity for wheat breeders to expand their attack on the winter hardiness barrier that has frustrated them for so long. Exploitation of this new technology to produce adapted, super-hardy winter wheat cultivars will require close cooperation between wheat breeders and biotechnologists. This interdisciplinary effort will be expensive and immediate break-throughs should not be expected.

Diseases and Insects

In western Canada, there are a large number of diseases

and insects that have the potential to cause economic losses in winter wheat. Plant resistance is the most environmentally safe, reliable, and cost effective means available for controlling these diseases and insects. However, existing winter wheat cultivars have poor resistance to most of the diseases and insects prevalent in this region. The poor resistance of western Canadian winter wheat cultivars has not been due to the absence of satisfactory sources of genetic resistance. Rather, the problem has been the lack of resources to mount sustained breeding programs that give the production of disease and insect resistant cultivars a high priority.

The high level of cultivar disease and insect resistance that is an accepted part of spring wheat production in western Canada was achieved only after many years of intensive, coordinated efforts among wheat breeders and pathologists. Similar levels of resistance could be incorporated into winter wheat. However, as was the case with spring wheat, the development of cultivars with a broad spectrum of disease and insect resistance can be expected to proceed slowly and often in small increments.

The efforts to develop rust resistant winter wheat cultivars provides an example of the difficulties winter wheat breeders have encountered in living up to the high standards of insect and disease resistance maintained by spring wheat breeders. The expansion of winter wheat out of its traditional production area in Alberta has focused considerable attention on the need for rust resistant cultivars. Rust does not overwinter on the Canadian prairies and each year must be reintroduced from the southern USA.

For most of western Canada, winter wheat reaches maturity before the rust inoculum has a chance to build up to significant levels. Since rust usually enters the southeast corner of the Canadian prairies first, the highest risk of damage exists in southern Manitoba and southeastern Saskatchewan. In the northern part of the USA Great Plains region, all the major winter wheat cultivars are susceptible to leaf rust and most have poor levels of stem rust resistance. Consequently, rust resistant cultivars adapted to this region are not available for use as parents in Canadian breeding programs.

So far efforts to transfer rust resistance from Canadian spring wheat cultivars to winter wheat have not been successful. In western Canada, Norwin and CDC Kestrel winter wheat cultivars have a slow rusting characteristic that delays the development of stem rust epidemics by a week to ten days. CDC Kestrel and breeding lines with this slow rusting reaction have had an average 40 percent yield advantage over Norstar in trials grown under favorable moisture and heavy, late rust epidemics in Saskatchewan and Manitoba.

However, even with a yield advantage of this magnitude in the presence of a rust epidemic, the requirement for a resistant rust rating has prevented the disease evaluation team of the Prairie Registration Recommending Committee for Grain (Figure 1) from "supporting" slow rusting winter wheat lines for registration as cultivars for production in western Canada.

Grain Yield

A farmers' income per acre (hectare) is determined by grain yield and price per bushel (tonne). The price that the farmer has received for No. 1 and 2 Canada Western Red Winter wheat has been similar to that of No. 3 Canada Western Red Spring wheat. Therefore, the top two premium wheat grades have been conceded to the

Canada Western Red Spring wheat class and the price premium paid for these grades must be compensated for by higher yields if the Canada Western Red Winter wheat class is to earn economic option for the farmer.

The price of Canada Western Red Winter wheat has been closer to that of the high yielding Canada Prairie Spring than the Canada Western Red Spring wheat class. Therefore, unless there is new evidence that price per bushel (tonne) can be increased to reflect superior quality, winter wheat breeders must comply with long term breeding objectives that will allow Canada Western Red Winter wheat to remain competitive in markets that are priced similarly to that of Canada Prairie Spring wheat. This can only be interpreted to mean that winter wheat breeding objectives must reflect the lower protein and higher yield targets of the Canada Prairie Spring class

As a class, Canada Prairie Spring wheat cultivars have a protein concentration that is two percent lower and a grain yield that is 20 percent higher than Canada Western Red Spring wheat cultivars. This reflects a trade off of 10 percent gain in grain yield for each percentage point sacrificed in grain protein concentration of spring wheat. Norstar winter wheat has had a 25 to 36 percent yield advantage over cultivars from the Canada Western Red Spring wheat class.

Release of the winter wheat cultivar CDC Kestrel in 1991 established that the high yield potential of semi-dwarf wheat can be combined with good winter hardiness to further raise the yield advantage held by winter wheat in western Canada. CDC Kestrel has maintained a yield similar to Norstar under high drought stress conditions. Under more favorable moisture conditions, CDC Kestrel has yielded 22 percent higher than Norstar. Given the realities of the marketplace, winter wheat breeders must place continued emphasis on consolidating and improving this yield advantage while maintaining their efforts to improve other important agronomic and quality traits.

Assessment of quality has always been important to the wheat industry. The first wheat quality evaluation tests were extremely simple by today's standards. The chew test for gluten development was considered a valuable measurement in the selection for bread quality by early breeders in this country. While use of this test has fallen by the wayside, visual assessments of characteristics like the physical appearance of the grain have remained important quality indicators.

Plump, sound, bright kernels are desired by millers for their high flour extraction rates and samples of dark vitreous kernels free of "piebald" or "yellowberry" are associated with high protein concentration. Minimum specific (bushel or hectolitre) weights are also still required in most wheat markets. Developments in the area of cereal chemistry soon established that wheat quality was much more complex than the early simple tests indicated.

However, identification of the basic components determining quality and an explanation of their mode of function and interrelationships have not proven to be an easy task. This absence of a clear definition of quality has resulted in a proliferation of tests, each of which is professed to measure some important quality property. To be useful in the screening of breeding material, a quality prediction test should be simple, inexpensive, rapid and require only a small sample size.

Measurements that can be made directly on the seed, such as kernel physical appearance, hectolitre weight, 1,000-kernel weight, kernel hardness, and grain protein concentration, fulfil these criteria best. Micro-milling and cooking tests (Table 2) have been developed. However, as one moves from measurements of the physical appearance of the grain to milling and baking, the tests become more complex, expensive, time consuming, and larger samples are required.

High milling quality is important for cultivar acceptance in the marketplace and, because complete milling evaluation requires large samples of grain and is very time consuming, a number of prediction tests have been developed to allow for the screening of lines from breeding programs (Table 2). Millers require wheat with high flour extraction rate because sale of flour realises a much greater profit than does bran. Flour extraction rate can be modified by mill adjustment. However, increasing the extraction rate increases the flour contamination with bran, which in turn affects the flour colour and ash content.

White flour colour is important in many products. Therefore, measures of flour colour and ash content are used to establish the upper limits for flour extraction in screening tests for milling quality.

Table 2. Characters that may be evaluated in the assessment of hard wheat quality.

Seed	*Mixing, dough development, and baking characteristics*
Bushel or hectolitre weight	Farinograph:
Kernel weight	Absortion
Kernel hardness (soft vs hard)	
Grain protein concentration	
	Arrival time
	Departure time

Contd....

Contd....

	Development time
	Stability
Milling	20 min. drop
	Tolerance index
	Valorimeter value
Flour yield (%)	
Flour ash (%)	Starch damage
Flour colour	Amylograph peak viscosity
	Wheat alpha-amylase
Mixing, dough development, and baking characteristics	Remix loaf volume
	Remix baking strength
Flour protein (%)	Remix absortion
Sedimentation	Remix crumb colour
Wet Gluten	Remix loaf appearance
Falling number	Canadian short process loaf volume
Viscosity	Canadian short process absortion
Mixograph:	Sponge-and-dough loaf volume
Development time	Sponge-and-dough absortion
Peak height	AACC loaf volume
Height @ 5 min.	Strength
Area	
Extensograph:	
Length	
Height	
Length/Height	
Area	

Loaf volume, cookie (or cake), and noodle measurements are usually considered the final quality measures of hard, soft, and durum wheat, respectively. For each of these quality tests there are a number of procedural modifications that have been developed to accentuate certain quality characters. For example, different flour blends, bromate levels and mixing times are used in the loaf volume test. These tests are valuable in the final stages of cultivar evaluation, when a complete description of quality is required. However, they require large samples of grain and the procedures are time consuming and costly. Therefore, as was the case for milling quality, a number of prediction tests have been developed to permit the screening of lines from breeding programs (Table 2).

Kernel Hardness

Bread and pastry products are two of the major end uses of wheat. Hard wheat is used in the production of bread and soft wheat in the production of pastry products. There are a number of simple kernel hardness tests that can be used to quickly segregate lines in wheat breeding programs into hard and soft wheat classes. Kernel hardness is also under simple genetic control making it one of the easiest quality characters to select for in a wheat breeding program.

Alpha-amylase

Alpha-amylase is an enzyme that degrades starch to sugars. High levels of this enzyme reduce the baking quality of wheat and, for this reason, many importing countries have set maximum allowable alpha-amylase levels. All grain contains some alpha-amylase, but its level quickly increases in mature grain that has been exposed to conditions favorable for sprouting. Weather conditions at harvest are a matter of chance; however, alpha-amylase levels in the seed can be kept to a minimum by the development of cultivars that have good sprouting resistance and are genetically low in this enzyme.

Most hard red spring wheat cultivars have been selected for a high level of sprouting resistance or post-harvest dormancy. High levels of seed dormancy are a mixed blessing in winter wheat because it is planted in the fall immediately after harvest. Consequently, a high level of post-harvest seed dormancy would require that winter wheat seed be held over for planting one year after harvest to ensure rapid, vigorous germination and plant establishment.

Livestock Feed

Wheat is used primarily as an energy source by livestock feeders. Digestibility (obtained from feeding trials) and gross energy content determine the digestible energy of a grain. Significant genetic variability has not been found for digestibility of wheat grain. Therefore, variability in the digestible energy of wheat is dependent upon differences in gross energy content. However, only limited variability has been reported in the gross energy content of wheat. Consequently, it is doubtful that meaningful increases in digestible energy of wheat could be achieved by breeding for improved gross energy content.

In addition, the kind and amount of variability in chemical constituents, such as crude fiber, ash, crude protein, and oil suggest that altering levels of these factors would not likely increase the gross energy content of wheat. Crude protein concentration and specific weight are factors in some feed wheat markets. Therefore, if large quantities of winter wheat were grown for feed, it is possible that low fiber, high protein feed cultivars would have a market advantage.

Protein is a primary quality component that influences most wheat grain quality measurements. Its importance to wheat quality is recognised in the marketplace and most exporting countries have some segregation of commercial grain lots on the basis of protein concentration. In hard wheats, the majority of the variation in loaf volume of bread can be attributed directly to differences in protein concentration.

Protein concentration can be measured quickly on small samples of grain. It is one of the least expensive quality tests to perform and, with modern technology, its determination need not be overly complex. However, where absolute measurements of protein concentration are required, care must be taken to insure accuracy of results. When differences in grain yield are corrected for, genetic variation in protein concentration is a maximum of two percentage points in wheat. The protein concentration of the winter wheat cultivar Norstar is in the middle of this range.

Consequently, cultivars with a one percent increase or a one percent decrease in protein concentration, relative to Norstar, represent the maximum influence that winter wheat breeders can expect to have on this important quality character. This compares to a potential 12 percentage point range attributable to environmental effects. As a result, protein concentration is the quality character least accessible to the plant breeder.

The direct impact that a winter wheat breeder can have in influencing protein concentration is restricted to the manipulation of the genetic factors that regulate the efficiency of nitrogen extraction from the soil, its translocation to the grain, and its use for the synthesis of protein. The complexity of nitrogen metabolism in the wheat plant suggests that many genes will have at least a small effect on protein concentration. A large number of genes with small effects and a large environmental influence makes the development of high yielding cultivars with high protein concentration a difficult breeding objective.

In the marketplace, grain protein concentration is considered independently of grain protein and grain yield. However, if viewed from the standpoint of the wheat plant, grain protein concentration is determined by the ratio of grain protein yield to total grain yield (grain protein concentration % = grain protein yield total grain yield X 100). For this reason, grain protein concentration is very dependent upon both grain yield and grain protein yield. This difference in viewpoint helps to explain some of the difficulties wheat breeders have to deal with when selecting for both high grain yield and high protein concentration.

Wheat protein contains approximately 17.5 percent nitrogen. This nitrogen is obtained from the soil. Therefore, available soil nitrogen has a direct influence on grain protein yield. Nitrogen uptake by the plant is determined by an unstable enzyme, nitrate reductase, that is regulated by the level of available soil nitrogen. Level of available soil nitrogen is controlled by environmental factors such as moisture, temperature, soil ion concentration, soil pH, and by the rate of nitrogen fertiliser applied by the farmer. Therefore, the significant effect that environment has on wheat protein yield should not be a surprise.

An evaluation of spring and winter wheat cultivars with a wide range of quality has shown that cultivar grain protein yield is closely related to total grain yield. In other words, one of the most effective ways of selecting for high grain protein yield in a winter wheat breeding program is to select for high grain yield. However, in most instances, the increase in total grain yield is greater than the increase in grain protein yield resulting in lower protein concentrations for higher yielding cultivars.

The negative relationship between grain protein concentration and grain yield has aggravated one of the deficiencies in the winter wheat cultivar CDC Kestrel. CDC kestrel has a protein concentration that averages 0.4 percent lower than Norstar when the available soil nitrogen requirements for maximum grain yield have been satisfied. However, this difference is magnified for N levels below those that produce maximum grain yield.

The control of available soil nitrogen levels is beyond the influence of winter wheat breeders. Unless their winter wheat is degraded to feed wheat, there is little monetary incentive for farmers to increase fertiliser rates to maintain quality standards because premiums are not paid for high protein concentration in the winter wheat class. Consequently, because wheat breeders can only manipulate protein concentration with great difficulty and within narrow limits, the Canadian cultivar development and marketing system has exerted continuing pressure on wheat breeders to sacrifice yield potential in breeding programs in order to maintain quality standards for grain protein concentration.

12

Intercropping Production Practices

Most grain-crop mixtures with similar ripening times cannot be machine-harvested to produce a marketable commodity, since few buyers purchase mixed grains. Because of limited harvest options with that type of intercropping, farmers are left with the options of hand harvesting, grazing crops in the field with animals, or harvesting the mixture for on-farm animal feed.

However, some intercropping schemes allow for staggered harvest dates that keep crop species separated. One example would be harvesting wheat that has been interplanted with soybeans, which are harvested later in the season. Another example is planting harvestable strips, also known as strip cropping. When two or more crops are growing together, each must have adequate space to maximise cooperation and minimise competition between them. To accomplish this, four things need to be considered:

1) spatial arrangement,
2) plant density,
3) maturity dates of the crops being grown, and
4) plant architecture.

Spatial Arrangement

There are at least four basic spatial arrangements used in intercropping. Most practical systems are variations of these.

— *Row intercropping*—growing two or more crops at the same time with at least one crop planted in rows.

— *Strip intercropping*—growing two or more crops together in strips wide enough to permit separate crop production using machines but close enough for the crops to interact.

— *Mixed intercropping*—growing two or more crops together in no distinct row arrangement.

— *Relay intercropping*—planting a second crop into a standing crop at a time when the standing crop is at its reproductive stage but before harvesting.

Plant Density

To optimise plant density, the seeding rate of each crop in the mixture is adjusted below its full rate. If full rates of each crop were planted, neither would yield well because of intense overcrowding. By reducing the seeding rates of each, the crops have a chance to yield well within the mixture. The challenge comes in knowing how much to reduce the seeding rates. For example, if you are planning to grow corn and cowpeas, and you want mostly peas and only a little corn, it would be easy to achieve this.

The corn-seeding rate would be drastically cut (by 80% or more), and the pea rate would be near normal. The field should produce near top yields of peas even from the lower planting rate and offer the advantage of corn plants for the pea vines to run on. If you wanted equal yields from both peas and corn, then the seeding rates would be adjusted to produce those equal yields.

Maturity Dates

Planting intercrops that feature staggered maturity dates or development periods takes advantage of variations in peak resource demands for nutrients, water, and sunlight. Having one crop mature before its companion crop lessens the competition between the two crops. An aggressive climbing bean may pull down corn or sorghum growing with it and lower the grain yield. Timing the planting of the aggressive bean may fix the problem if the corn can be harvested before the bean begins to climb.

A common practice in the old southern U.S. cotton culture was to plant velvet beans or cowpeas into standing corn at last corn cultivation. The corn was planted on wide 40-inch rows at a low plant population, allowing enough sunlight to reach the peas or beans. The corn was close enough to maturity that the young legumes did not compete. When the corn was mature, the beans or peas had corn stalks to climb on. The end result was corn and beans that would be hand harvested together in the fall. Following corn and pea harvest, cattle and hogs would be turned into the field to consume the crop fodder.

Selecting crops or varieties with different maturity dates can also assist staggered harvesting and separation of grain commodities. In the traditional sorghum/ pigeonpea intercrop, common in India, the sorghum dominates the early stages of growth and matures in about four months. Following harvest of the sorghum, the pigeonpea flowers and ripens. The slow-growing pigeonpea has virtually no effect on the sorghum yield.

Plant Architecture

Plant architecture is a commonly used strategy to allow one member of the mix to capture sunlight that would not otherwise be available to the others. Widely spaced corn plants growing above an understory of beans and pumpkins is a classic example.

Intercrop Productivity

One of the most important reasons to grow two or more crops together is the increase in productivity per unit of land. Researchers have designed a method for assessing intercrop performance as compared to pure stand yields. In research trials, they grow mixtures and pure stands in separate plots. Yields from the pure stands, and from each separate crop from within the mixture, are measured. From these yields, an assessment of the land requirements per unit of yield can be determined. This information tells them the yield advantage the intercrop has over the pure stand, if any. They then know how much additional yield is required in the pure stand to equal the amount of yield achieved in the intercrop.

The calculated figure is called the Land Equivalency Ratio (LER). To calculate an LER, the intercrop yields are divided by the pure stand yields for each component crop in the intercrop. Then, these two figures are added together. Here's the equation for a corn/pea intercrop where the yields from pure corn, pure peas, and the yields from both corn and peas growing together in an intercrop are measured.

$$(\text{intercrop corn} \;/\; \text{pure corn}) + (\text{intercrop pea} \;/\; \text{pure pea}) = \text{LER}$$

When an LER measures 1.0, it tells us that the amount of land required for peas and corn grown together is the same as that for peas and corn grown in pure stand (i.e., there was no advantage to intercropping over pure stands). LERs above 1.0 show an advantage to intercropping, while numbers below 1.0 show a disadvantage to intercropping. For example, an LER of 1.25 tells us that the yield produced in the total intercrop would have required 25% more land if planted in pure stands.

If the LER was 0.75, then we know the intercrop yield was only 75% of that of the same amount of land that grew pure stands. In a South Carolina study, researchers planted intercrops of southern peas and sweet corn at three different corn plant densities. The plantings were on raised beds with flat and wide crowns on six-foot centers. In the center of each bed was a corn row, with two rows of peas planted 18 inches to either side of the corn row (Figure 1).

The low corn-seeding rate was 6,700 plants per acre, medium corn was 9,500 per acre, and high was 11,900 plants per acre. Peas were established at a rate of 31,800 plants per acre in all intercrop plots. In the pure pea stand, each bed had two rows of peas spaced 24 inches apart. Yields of the intercrops and pure stands are shown in Table 1. In this trial there was a yield advantage from intercropping over growing the two crops in pure stands. Pea yields suffered from the increased competition in the higher densities of corn.

Some practical on-farm guidelines can be drawn to guide seeding-rate choices for a two-crop intercrop. To test seeding rates, experiment with three small plantings of two crops at the following percentages of their full seeding rates: 1/3 + 2/3, 1/2 + 1/2, and 2/3 + 1/3. From there, make adjustments for future plantings based on the results and your expectations.

Table 1. Yields of sweet corn and southern peas from intercrops.

Seed Rates	*Corn (pounds/acre)*	*Peas (pounds/acre)*	*LER*
Full corn	5600	***	***
Full peas	***	1200	***
Low corn	4200	800	1.41
Medium corn	4600	800	1.48
High corn	5000	500	1.30

Intercrops Management

Many combinations of crops have been grown or experimented with as mixed or relay intercrops. Some of these include sunflowers grown with black lentils, wheat with flax, and canola with flax. Other combinations include cucumbers, beans, celery, and chives in China; upland rice, corn, and cassava in Indonesia, and in various parts of the tropics corn and cassava, corn and peanut, sorghum and millet, and sorghum and pigeonpeas.

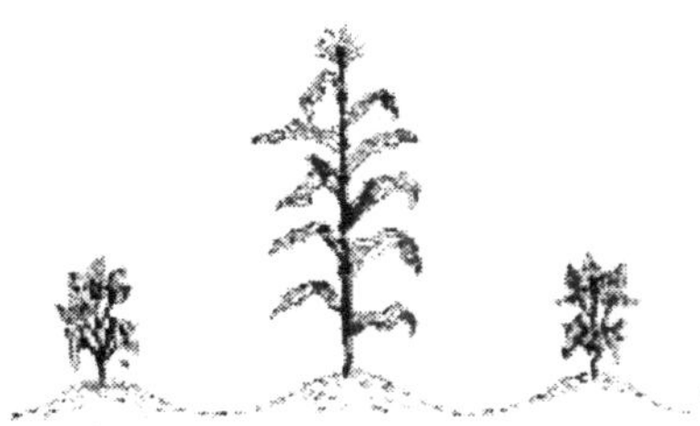

Figure 1. Sweetcorn and southern pea planting pattern

Frequently these cropping combinations involve a short and a tall crop both planted at the same time. In many cases the tall crop is harvested first. For example, corn grown with a shorter plant would be harvested first, then peanut or sweet potato would be harvested later. Another pattern would be planting two tall crops with different growth rates. In relay intercrops, different planting dates are used so that one crop might mature sooner. Corn or sorghum, requiring three months to mature, can be grown with pigeonpea, requiring 10 months to maturation.

John Bowen and Bernard Kratky, researchers and instructors at the University of Hawaii, tell us that there are five distinct aspects to successful multiple cropping. These are:

1) detailed planning,
2) timely planting of each crop,
3) adequate fertilisation at the optimal times,
4) effective weed and pest control,
5) efficient harvesting.

Before any fieldwork is begun, adequate planning should be done. Planning covers selection of crop species and appropriate cultivars, water availability, plant populations and spacing, labour requirements throughout the season, tillage requirements, and predicted profitability of the intercrop. These and other parameters need to be evaluated before spending money on inputs. With any crop, seed germination and seedling establishment are the most critical phases of the entire season.

A good seedbed is needed to get a good stand. Delayed planting may reduce yield, since crop development may not coincide with the optimal growth periods. Planning fertilisation for intercrops can be challenging, as the full needs of both crops must be met.

Generally, there is little information available on how to go about this. One possibility would be to ask for soil test results for each crop separately, then formulate a recommendation that will cover the needs of both crops to be grown. Such recommendations are generally 10% to 30% higher than rates for individual crops. As with any crop, also accounting for residual or carryover fertility from past crops saves money.

Carryover fertility from intercrops may well be lower than that of pure stands because of the two crops having different root types and feeding habits. Weed and pest controls need in intercrops will likely be different from those in pure stands. Some disease incidence, such as soybean or mung bean rusts, may increase when aggravated with high corn populations and overfertilisation. Any disease or pest that prospers in shady conditions could increase under a taller crop such as corn or sunflowers. In many cases, insect pest populations are lower when two or more crops are grown together.

Harvesting of mixed intercrops has been a major limitation to their adoption in mechanised farming. As mentioned earlier, if the crops cannot be harvested by animals, or all together as feed, you're left with hand harvesting. Some crops such as flax and wheat have been harvested together and mechanically separated. Any other mechanised harvest efforts must get one crop without damaging the other. One example would be harvesting wheat over the top of a young stand of soybeans growing beneath the grain heads. All intercropping strategies- especially mixed intercropping- require advanced planning and keen management. Success will likely be the reward for such efforts.

Examples of Intercropping

Traditional corn-bean-squash mixed intercrops

Farmers throughout Central America traditionally grow an intercrop of corn, beans and squash. Grown together, these three crops optimise available resources. The corn towers high over the other two crops, and the beans climb up the corn stalks. The squash plants sprawl along the ground, capturing light that filters down through the canopy and shading the ground. The shading discourages weeds from growing. This mixture was compared to the individual crops grown separately in a study near Tabasco, Mexico.

In the study, corn yields were considerably higher in the mixture than in a pure stand planted at optimum densities. Bean and squash yields suffered considerable yield reductions when grown in mixture. In this example if corn were the most important crop, it was beneficial to grow it in a mixture with squash and beans. The beans and squash were just a bonus. The LER for the whole mixture was considerably higher than any of the pure stands.

Table 2. Yields of corn, beans and suash grown alone or in a mixture

Crop	*Pure Stand (pounds/acre)*	*Intercrop (pounds/acre)*
Corn	1096	1533
Beans	544	98
Squash	383	71

Corn and soybean mixed intercrops

Canadian researchers have worked with several corn-soybean intercrop seeding rates to determine their economic advantages as silage. Pure stands of corn and soybeans were grown for comparison at 24,000 corn seed per acre and 200,000 soybean seed per acre. Results showed that intercrops were more cost effective than pure stands over both years the study was conducted. The study featured five experimental intercrop seeding rates with two planting arrangements (alternate and within the row).

The researchers concluded that a planting rate of 16,000 corn seed per acre (67% of the full corn rate) with 135,000 soybean seed per acre (67% of the full bean rate) planted within the same rows along with 53 lbs. of N/acre gave the highest economic returns. (Note: the planter was set to drop 151,000 seeds per acre.) This mixture gave an LER of 1.14 over pure stand yields. The crude protein level of the intercrop silage was considerably higher than that of pure corn silage. A slightly higher yield was achieved from full stands of both corn and beans in alternate rows (LER=1.23), but the cost of production was higher, thus offsetting the improved yields.

Corn and sorghum mixed intercrops

Frank Cawrse, Jr., of Lebanon, Oregon, intercrops forage sorghum into his silage corn. He first plants the corn at 28,000 seed per acre, then goes back over the field with a drill with enough drop tubes closed off to plant 8 pounds of sorghum on 32-inch rows in between the corn. He also plants two different maturities of corn, a 95-day and a 75-day, to even out the silage moisture content. He harvests a mix of corn in hard dent and soft dent, and sorghum in the milk stage.

Strip cropping corn/soybeans/small grains

South Dakota farmer Tod Intermill plants alternating strips of corn, soybeans, and spring wheat on his farm.

The strips are six rows wide in a ridge-till system. All the crop plantings are adapted to existing equipment widths. Regular herbicide treatments can be applied using a ground sprayer of strip width. Even the wheat is drilled on ridges, using a drill with individual depth gauges on each opener.

Intermill orients his rows east and west to minimise the shading effects of taller crops like corn. The crops are planted in a wheat-corn-soybean pattern, with soybeans on the north side of the corn. This arrangement reduces the effect of corn shading often associated with a straight corn-soybean pattern, since the wheat is mature before the corn has a chance to shade it. Corn gains the greatest benefit from the additional sunlight interception on the outside rows of the corn strip.

Iowa farmer Tom Frantzen strip-crops oats, corn, and soybeans on ridge-till rows. He views his strips as a crop rotation in one field. His rows are oriented generally east and west on the contour. His 1989 strip-crop corn yields were 166 bushels per acre, compared to 130 for his farm average. Stripped soybean yields were two bushels lower than farm average. His oat yields were 109 bushels stripped and 100-bushel farm average. Tom was not surprised at the increase in corn yields.

The outer strip rows captured more sunlight. His average corn border row yielded 198 bushels per acre next to the soybeans and 177 bushels next to oats. The soybean yields were 37 bushels, even with the increased shading on the border rows. This loss was made up in the middle rows with yields of 44 bushels per acre. Oats showed a 107-bushel yield on the soybean side, a 103-bushel yield on the corn side, and 99 bushels in the middle. Tom says the strip intercropping is no more labour intensive than monocrop fields. His profits were

$76 per acre for the stripped fields and $55 for the same crops grown in monoculture.

Rick Cruse, an Iowa State University agronomist, has observed several characteristics that narrow strips (12 to 30 feet wide) offer. The strips accommodate the pest management and soil building advantages of rotations and the yield boost of border rows. With proper management the border effect can pay off; managed improperly, it can cost yield. With oat and corn strips, the early-maturing oats are nearly mature before corn can pose much of a shade and competition problem.

The corn can also provide wind protection for the oats. When the oats are harvested, more sunlight is available to the corn. In times of low moisture, oats may rob the corn border rows of water. In his field trials, Cruse found a 5% increase in oat yields on their borders while corn realised a 12 to 15% increase. Soybean yields dropped by 10% on their border rows, but the yields in the soybean middle rows were higher than they would be in a solid field, possibly representing a windbreak effect.

Some have experimented with a shorter corn variety in the border row to minimise shading. One farmer tried planting six rows of corn and doubling his soybean strips to 12 rows to eliminate the impact of corn shading on the beans. This same farmer found that corn strips wider than eight rows did not provide adequate results. Using a 12-row planter, it's easy to establish the 6-row strips by filling the middle six hoppers with corn and the outer three hoppers with beans.

Some farmers plant higher corn populations and add higher nitrogen rates in the border rows to take advantage of the extra sunlight exposure. Most farmers agree that strip cropping corn, soybeans, and oats works best with ridge-till or no-till. When the field is tilled, it's difficult to gauge where the rows should go in order to get the strips even.

Diversity and Stability of Natural Systems

Ecologists tell us that stable natural systems are typically diverse, containing many different types of plants, arthropods, mammals, birds, and microorganisms. In stable systems, serious pest outbreaks are rare, because natural controls exist to automatically bring populations back into balance. Planting crop mixtures, which increase farmscape biodiversity, can make crop ecosystems more stable, and thereby reduce pest problems.

There is overwhelming evidence that plant mixtures support lower numbers of pests than do pure stands and there are two schools of thought on why this occurs. One suggests that higher natural enemy populations persist in diverse mixtures due to more continuous food sources (nectar, pollen, and prey) and favorable habitat.

The other thought is that pest insects that feed on only one type of plant have greater opportunity to feed, move around in, and breed in pure crop stands because their resources are more concentrated than they would be in a crop mixture. Regardless of which reason you accept, the crops growing together in the mixture complement one another, resulting in lower pest levels.

Intercropping also aids pest control efforts by reducing the ability of the pest insects to recognise their host plants. For example, thrips and white flies are attracted to green plants with a brown (soil) background, ignoring areas where vegetation cover is complete-including mulched soil. Some intercrops have a spatial arrangement that produces the complete vegetation cover that would be recognised as unfavourable to thrips and whiteflies. Other insects recognise their host plants by smell. Onions planted with carrots mask the smell of carrots from carrot flies.

Innovative farmers are paving the way with intercrops and realising pest management benefits as a result. Georgia cotton farmers Wayne Parramore and sons reduced their insecticide and fertiliser use by growing a lupine cover crop ahead of their spring-planted cotton. They started experimenting with lupines on 100 acres in 1993, and by 1995 were growing 1,100 acres of lupines. Ground preparation for cotton planting is begun about 10 days prior to planting by tilling 14-inch wide strips into the lupines. Herbicides are applied to the strips at that time, and row middles remain untouched.

The remaining lupines provides beneficial insect habitat and also serve as a smother crop to curtail weeds and grasses. The lupines in the row middles can be tilled in with the cultivator later in the season to release more legume nitrogen.

In the Parramores' system, all the nitrogen needs of the cotton crop are met with cover crops except for 10 units per acre of starter nitrogen and another 15 units applied while spraying herbicides. Petiole samples taken every week to monitor plant nitrogen show that cotton grown with lupines maintains a normal range of tissue nitrogen throughout the growing season. The nitrogen level in cotton grown solely with fertiliser is very high initially, then subsequently falls back to a lower level.

In one comparative year, the cotton grown following lupine produced 96 more pounds of lint, with only 25 units of commercial nitrogen, compared to a field with 125 units of nitrogen and no lupines. Additionally, the lupine field required less spraying for insects-only twice compared to five sprays for the commercial nitrogen field. This reduction saved 60% on insecticides, amounting to $35 per acre. The reduction in need for pesticides is attributed to the large population of beneficial insects generated and sustained in this system.

The lupines provide food for aphids and thrips, which attract ladybugs, big-eyed bugs, and fire ants as predators. When the cotton gets big enough to shade out the lupines, the beneficial insects move to the cotton rather than migrating from the field. The Parramores estimate that improved yields, combined with cost reductions, are netting them an additional $184 per acre with the strip tillage lupine system when compared to the conventional management system.

Alfalfa is one of the best crops for attracting and retaining beneficial insects. This characteristic can be enhanced further. Strip-cutting alfalfa (i.e., cutting only half of the crop at any one time, in alternating strips) maintains two growth stages in the crop; consequently, some beneficial habitat is available at all. times. In some cases alfalfa is mixed with another legume and a grass. Auburn University researcher Mike Gayler is just starting research projects using alfalfa as an attractant crop for beneficials. He speculates that it will work in the Southeast with proper management. Other main-season strip crops that research suggests will benefit cotton crop pest management include cowpeas, sorghum, corn, and crotalaria.

Dr. Sharad Phatak of the University of Georgia has been working with cotton growers in Georgia testing a strip-cropping method using annual winter cover crops. Planting cotton into strip-killed crimson clover improves soil health, cuts tillage costs, and allows him to grow cotton with no insecticides and only 30 pounds of nitrogen fertiliser. Working with Phatak, farmer Benny Johnson reportedly saved at least $120/acre on his 16-acre test plot with the clover system.

There were no insect problems in the test plot, while beet armyworms and whiteflies were infesting nearby cotton and requiring 8 to 12 sprayings to control. Cotton intercropped with crimson clover yielded more than three bales of lint per acre compared to 1.2 bales of lint per acre in the rest of the field. Boll counts were 30 per plant with crimson clover and 11 without it. Phatak identified up to 15 different kinds of beneficial insects in these strip-planted plots.

Phatak finds that planting crimson clover seed at 15 pounds per acre in the fall produces around 60 pounds of nitrogen per acre by spring. By late spring, beneficial insects are active in the clover. At that time, 6- to 12-inch planting strips of clover are killed with Roundup™ herbicide. Fifteen to 20 days later the strips are lightly tilled and cotton is planted. The clover in the row-middles is left growing to maintain beneficial insect habitat.

When the clover is past the bloom stage and less desirable for beneficials, they move readily onto the cotton. Even early-season thrips, which can be a problem following cover crops, are limited or prevented by beneficial insects in this system. The timing coincides with a period when cotton is most vulnerable to insect pests. Following cotton defoliation, the beneficials hibernate in adjacent non-crop areas. Phatak points out that switching to a whole-farm focus while reducing off-farm inputs is not simple. It requires planning, management, and several years to implement on a large scale. It is just as important to increase and maintain organic matter, which stimulates beneficial soil microorganisms. Eventually a "living soil" will keep harmful nematodes and soilborne fungi under control.

Texas dryland farmer Ron Gobel intercrops 8-row strips of sesame and cotton for insect control benefits. The sesame harbors many beneficial insects, including high populations of lace-wings, assassin bugs, and lady beetles. Ron's 1995 crop was planted late due to prolonged spring rains. He did not use a Bt cotton variety. Early frost terminated the crop two weeks earlier than normal yet he still produced 0.8 bales per acre under dryland conditions. His sesame produced 800 pounds per acre. The 1996 cotton rows were planted where the sesame rows were the previous year, and sesame planted where cotton was before.

Since Ron sells his cotton for a premium price in the organic market, he cannot spray any synthetic insecticides. Consequently, he must rely on beneficial insects attracted to his fields by cultural practices and a handful of natural insecticides. Following the fall harvest Ron plants annual rye at a low rate of 20 to 40 pounds per acre. The rye is tilled in prior to crop planting in the spring. Ron believes the rye helps with soil moisture retention and weed control.

During the 1997 crop year his fields suffered only minimal boll weevil damage. Ron noticed lots of adult bollworm moths but no worms. The eggs were eaten or parasitised by the beneficials. Ron's fields were scouted as part of a boll weevil eradication program. The scouts were amazed at the lack of worms and the high numbers of beneficial insects. The cotton crop was sprayed one time with diatomaceous earth impregnated with natural pyrethrum, which was acceptable under the organic standards.

The insect scouts noticed a 70% reduction in adult boll weevil population three days after the spray. They were so surprised, that they placed cages of 20 live weevils in the field to see whether the spray was working. The next day, 45% of those weevils were dead. The entomologists speculated that the weevils were getting enough of the diatomaceous earth on their leg joints to cut their exoskeletons, allowing the pyrethrum to kill them. In a scientific study, Mississippi researchers interplanted 24 rows of cotton with 4 rows of sesame to study the intercrop's effects on tobacco budworms and bollworms (*Heliothis spp.*).

Throughout the growing season, larvae numbers were much higher in the sesame than on the cotton until late August, indicating the worm's preference for sesame. Following a large summer rain at a time when the sesame was reaching maturity, the Heliothis adults became more attracted to the cotton. The researchers noted that sesame's attractiveness to Heliothis and sesame's ability to harbour high numbers of beneficial insects made it useful in a cotton pest management program.

The diversity created by intercropping can be enhanced even further by integrating livestock (single or mixed species) into the cropping plan as harvesters. Allowing animals to harvest feed crops in the field puts gain on animals at the cost of crop production-considerably less than the purchase price of the grain. If you think about it, feed grains cost a lot less when they're not run through a $150,000 combine or hauled 1000 miles across the country. Grazing animals and other livestock can be managed on croplands to reduce costs, increase income, and increase diversity. There are ways of incorporating animals into cropping without the farmer getting into animal husbandry or ownership directly.

Collaboration with neighbours who own animals will benefit both croppers and livestock owners. Grazing or hogging-off of corn residue is one example where a cost can be turned into a profit. The animals replace the $6 per acre stalk mowing cost and produce income in animal gains. Shasta College provides a unique demonstration of integrating livestock with intercrops. Shasta is a two-year community college located in Redding, California, that offers associate degrees in several branches of agriculture.

Stan Gorden heads the college's holistic resource laboratory, where students get hands-on experience with ranching and farming. Stan and his students have taken intercropping to a high level of efficiency. They run hogs, sheep, cattle, and chickens together over 42 small paddocks of various forages and crops growing on 100 acres of college-owned land. One paddock is a pumpkin patch, another a garlic and carrot patch. Some are planted in alfalfa or mixes of grasses and clover. Not all the pastures have water sources for the animals, so water is moved on a trailer tank when necessary. The animals are moved daily in a planned grazing system during rapid plant growth and much more slowly, up to five days on a paddock, during slow plant growth.

Some of the paddocks are planted with mixtures of either winter or summer forage or grain crops. An intercrop of cereal grain, fava beans, and Canadian field peas is planted for winter grain, each crop at 1/3 normal seeding rate. The grain mixture is combine-harvested to make energy and protein supplement feed as needed. After harvest, the animals are turned into the paddock to glean what's left. For summer feed, a mixture of milo planted on 18-inch rows is intercropped with a row of black-eyed peas planted six inches to either side of each sorghum row, using a drill with partitions in the seedbox.

The milo provides a trellis for the pea vines to run on (Figure 2). The milo/black-eyed mixture requires no herbicide. Before peas and milo were grown together, the milo pure stand would be plagued with whiteflies and green bugs. Mixing the two crops together ended the pest problem. Cowpeas have extrafloral nectaries that attract lots of beneficial insects. This could explain the absence of pest insects in the mixture. The milo/pea mixture is harvested by setting the combine to cut at the height of the milo heads. This yields a milo to bean ratio of 2:1-ideal for feed.

Figure 2. Cowpeas and milo growing together

The college animal herd consists of 20 sows that farrow on pasture, 35 head of cattle, 50 sheep, and 30 laying hens that all range together. The hens are with the herd during the day and roost in a nearby eggmobile at night. Gorden selects breeds and genetics to fit this system, as opposed to selecting breeds for maximum production, and adapting a system to match the animal. The animals benefit one another. The sheep learn to stay close to the middle of the herd to avoid predators, which are fended off by the hogs.

The cattle learn that the hogs know how to break the pumpkins open, so they stick close and get some too. The hogs eat the cow and sheep droppings and benefit from the predigestion. The hens scavenge wasted seeds from the various crops. There are three different kinds of hens, each of which lays eggs of a different colour. The eggs are marketed as rainbow eggs, with each dozen containing four white, four blue, and four brown eggs. The chickens also scratch apart cattle dung pats searching for insects, thus destroying cattle parasites.

Developing and maintaining this high level of diversity has required creativity, selection criteria, constant monitoring, and re-examining traditional beliefs. By challenging long-held beliefs, he and his students discovered that hogs do not need standard farrowing crates and that sheep and cattle are compatible grazers. Animals and crops are selected and culled accoıding to their ability to adapt to this complex system.

The hogs have been fitted with humane nose rings to prevent rooting. Also, hog breeds are selected that don't root up the ground nor eat the baby lambs when they are born. The sows farrow on pasture with only a single bale of hay for bedding. Hogs are not vaccinated, nor are needle teeth removed or other detailing done. Sows generally wean 12 pigs with no supplemental feed. The only purchased input is some nitrogen and phosphorus fertiliser applied to the pastures. The pigs are only touched twice; once to castrate and once to wean.

As with the hogs, the cattle and sheep are selected to prosper on grass. Predators are not controlled in any way. Any animal that gets killed by wandering off is naturally selected out of the herd. The sheep/hog/cow mix provides much better utilisation of forage than single species grazing. Since the animals do most of the harvesting, less fossil fuel and labour-hours are expended. There are no pens to wash and no manure to deal with.

Use of Intercropping

Under direction of an international team of scientists, farmers in China's Yunnan province made some simple changes in their rice production methods. They changed from planting their typical pure stand of a single rice variety to planting a mixture of two different rice varieties. Their primary reason for trying this new technique was to reduce the incidence of rice blast, the main disease of rice. The technique was so successful at reducing blast disease that the farmers were able to abandon chemical fungicides they had been using.

The biodiversity effect is apparent here in that if one variety of a crop is susceptible to a disease, the denser the stand, the worse the disease can spread. If susceptible plants are separated by non-host plants that can act as a physical barrier to the disease, the susceptible variety will suffer less disease infection. Rice blast moves from plant to plant via airborne spores. These spores can be blocked by a row of a resistant variety. In this on-farm study, the rice was harvested by hand. Separating the varieties was easily done during harvest, since one variety towered above the other.

Intercropping has been important in the U.S. and other countries and continues to be an important practice in developing nations. In traditional systems, intercropping evolved through many centuries of trial and error. To have persisted, intercropping had to have merit biologically, environmentally, economically, and sociologically. To gain acceptance, any agricultural practice must provide advantages over other available options in the eyes of the practitioner.

Many of the impediments to adoption of new strategies or practices of diversification are sociological and financial rather than technological. Farmers have generally regarded intercropping as a technique that reduces risks in crop production; if one member of an intercrop fails, the other survives and compensates in yield to some extent, allowing the farmer an acceptable harvest. Pest levels are often lowered in intercrops, as the diversity of plants hampers movement of certain pest insects and in some cases encourages beneficial insect populations.

13

Crop Breeding for Market Orientation

Molecular marker-assisted selection, often simply referred to as marker-assisted selection (MAS) involves selection of plants carrying genomic regions that are involved in the expression of traits of interest through molecular markers. With the development and availability of an array of molecular markers and dense molecular genetic maps in crop plants, MAS has become possible for traits both governed by major genes as well as quantitative trait loci (QTLs). A generalized and simplified key for distinction between qualitative and quantitative traitsis presented in Table 1.

If the individual gene(s) or QTLs with significant influence on specific target trait(s) can be identified based on their linkage to molecular makers, the efficiency of incorporating the desired traits in elite germplasm could be greatly enhanced. While plenty of information is already available on different kinds of marker systems and gene-marker associations, the practicalities of designing an MAS strategy, increasing their chances of success and efficiency have not received adequate attention till now. A number of molecular investigations

Table 1. Qualitative vs quantitative traits

Trait	Segregation	Percentage of phenotypic variation explained	Example	Classification
Qualitative	Discrete	100	Purple leaf colour, blast resistance in rice, anthocyanin pigmentation and opaque endosperm in maize, etc.	Major gene
Semi-quantitative	Discrete	100	Semidwarfism in rice, *sd1* etc.	Major gene
Quantitative	Continuous	> 50	Submergence tolerance gene *sub1* in rice, QTLs for most biochemical traits, etc.	Major gene
Quantitative	Continuous	25–50	Stem rot resistance in rice, etc.	Major QTL
Quantitative	Continuous	< 25 QTL	for most agronomic and physiological traits	QTL

have revealed useful gene-marker associations in different crops, but routine implementation in ongoing plant breeding programmes is still in its infancy. It is an opportune time to discuss the practical aspects of designing an MAS strategy and high throughput genotyping techniques that increase their efficiency significantly.

In general, the success of a marker-based breeding system depends on three main factors: (i) A genetic map with an adequate number of uniformly-spaced polymorphic markers to accurately locate desired QTLs or major gene(s); (ii) Close linkage between the QTL or a major gene of interest and adjacent markers; (iii) Adequate recombination between the markers and rest of the genome; and (iv) An ability to analyse a larger number of plants in a time- and cost- effective manner.

The success of MAS depends on location of the markers with respect to genes of interest. Three kinds of relationships between the markers and respective genes could be distinguished;

i) The molecular marker is located within the gene of interest, which is the most favourable situation for MAS and in this case, it could be ideally referred to as gene-assisted selection. While this kind of relationship is the most preferred one, it is also difficult to find this kind of markers. For instance, microsatellite or simple sequence repeat (SSR) markers have been designed using the available DNA sequence information for the *opaque2* allele that confers high lysine and tryptophan content in the maize kernel. This has offered an efficient means of tracking the *opaque2* allele in breeding for nutritionally superior maize genotypes, since the marker is located within the gene sequence itself and cosegregates with the target gene.

ii) The marker is in linkage disequilibrium (LD) with the gene of interest throughout the population. LD is the tendency of certain combination of alleles to be inherited together. Population-wide LD can be found when markers and genes of interest are physically close to each other. Selection using these markers can be called as LD-MAS.

iii) The marker is in linkage equilibrium (LE) with the gene of interest throughout the population, which is the most difficult and challenging situation for applying MAS.

However, in most cases, especially for the polygenic traits, the target gene(s) within a QTL have not been characterized at the molecular level. Therefore, genomic regions to be selected using MAS are often chromosome segments carrying QTLs in case of polygenic traits. It is preferable either to have two polymorphic DNA markers flanking the target gene (or a QTL), or a marker within a QTL (if the chromosome segment is more than 20 cM) to eliminate the possibility of genotypes presenting a double recombination between the two flanking markers. Depending on the nature of the genomic region (cloned gene, major or minor QTL) involved in the expression of a target trait and the number of selected QTLs or genomic regions that need to be manipulated, several MAS schemes have been proposed that will be discussed later.

In the context of MAS, DNA-based markers can be effectively utilized for two basic purposes: (i) tracing favourable allele(s) (dominant or recessive) across generations and (ii) identifying the most suitable individual(s) among the segregating progeny, based on allelic composition across a part or the entire genome.

Foreground and Background Selection

Molecular markers are now increasingly being employed to trace the presence of target genes (foreground selection) as well as for accelerating the recovery of the recurrent parent genome (background selection) in backcross pro-grammes. Marker-assisted backcrossing (MAB) improves the efficiency of backcross breeding in three ways:

(i) If the phenotype of the desired gene cannot be easily assayed, backcross (BC) progeny possessing a marker allele from the donor parent at a locus near/within the target gene can be selected with a good probability of carrying the gene;

ii) Markers can be used to select BC progeny with least amounts of donor parent germplasm in the genome outside the target region; and

iii) Markers can be used to select rare progeny that are the result of recombination near the target gene, thus minimizing the effects of linkage drag.

Transfer of recessive genes through conventional breeding requires additional selfing generations after every backcross, a procedure that is prohibitively slow for most commercial breeding purposes. Marker-assisted foreground selection was effectively utilized for introgression of disease resistance genes by Melchinger, who presented an a priori approach for calculating the minimum number of individuals and family size required in recurrent backcrossing. However, due to lack of allele-specific markers, practical examples of this approach in plant breeding are limited. One successful example is the conversion of normal maize lines into quality protein maize (QPM) through marker-assisted transfer of a recessive mutant allele, *opaque2*, using allele-specific molecular markers. In animal breeding, availability of an array of allele-specific markers has been facilitating applications of this approach on a commercial scale to eliminate disease and stress-susceptibility genes.

'Marker-assisted background selection', a term coined by Hospital and Charcosset, was initially proposed by Young and Tanksley, and experimented by various researchers. This strategy has been used extensively in commercial maize breeding programmes, particularly for selection of lines carrying transgenes conferring herbicide tolerance or insect resistance. Several parameters need to be optimized in the background selection programmes. Flanking markers for the target allele are necessary to remove linkage drag. The optimal distance between the target gene and flanking markers governs the selection intensity that can be exerted. The equations given by Hospital et al.and Frisch et al.are helpful in determining the number of BC plants that need to be generated and typed with a special set of flanking markers.

Improvement of Qualitative Traits

When the expression of a target trait is controlled by a single gene or by a gene that accounts for a high proportion of the phenotypic variance of the trait, transfer of that specific gene from the donor to the recipient line can lead to significant improvement of the trait. For the introgression of qualitative traits such as pathotype-specific disease resistance or certain quality characteristics, which are typically governed by single dominant/recessive genes, backcross breeding has been used for a long time.

Traditional backcrossing programmes are planned on the assumption that the proportion of the recurrent parent genome is recovered at a rate of 1 - (1/2)for each of t generations of backcrossing. Thus, after four backcrosses, we expect to recover 1 - (1/2)= 96.9% of the recurrent parent genome. However, any specific BC progeny will deviate from this expectation due to chance and to linkage between the gene from the donor parent being selected for and nearby genes. A good example of surprising amount of the linkage drag that accompanies backcross breeding programmes was reported by Young and Tanksley, who genotyped the chromosome carrying the *Tm2* disease resistance gene in several tomato cultivars that were developed by introgressing the gene from a wild relative, Lycopersicon peruvianum via backcross breeding.

They found that even cultivars developed after 20 backcrosses contained introgressed segments as large as 4 cM and one cultivar developed after 11 backcrosses still contained the entire chromosome arm carrying the gene from the donor parent. Besides, for the transfer of a single dominant gene, a minimum of six backcross generations would normally be required to recover 99% of the recurrent parent genome. This procedure is too time-consuming, particularly in the context of the competitive nature of modern hybrid breeding programmes, where the turnover times for new lines and hybrids are fast.

There are three of the successful cases of marker-aided backcrossing intended towards the improvement of qualitative traits.

MAS for resistance to soybean cyst nematode: Diseaseresistant phenotypes are often simple and oligogenic in nature. Yet, the difficulties in establishing reliable inoculation and scoring methods can challenge even the best plant pathologist or breeder. An example is the resistance to soybean cyst nematode (SCN; Heterodera glycines). A widely used phenotypic assay takes five weeks, extensive greenhouse space and 5-10 h of labour for every 100 plant samples processed. The need of the hour for research groups was to develop tightly linked and breederfriendly markers for the major SCN resistance genes. Due to consistent research efforts, microsatellite or SSR markers have become available for the rhg1 gene, which is primarily responsible for SCN resistance.

Currently, the SSR marker sat309 located only 1-2 cM away from the rhg1 forms the basis of most public breeding efforts. This marker alone can be extremely effective in screening breeding populations. Genotypic selections with sat309 were 99% accurate in predicting lines that were susceptible in subsequent greenhouse assays. Recently, high throughput genotyping using Taqman??probes for a polymorphism linked to SCN resistance gene, Rhg4 has been standardized, which is discussed in detail later in this review.

MAS in developing QPM genotypes: Normal maize protein is inherently deficient in two essential amino acids, namely lysine and tryptophan. QPM refers to maize genotypes in which the *opaque2* gene is introgressed along with endosperm modifiers that provide the hard kernel texture. QPM, thereby, offers twice the amount of lysine and tryptophan with kernel vitreousness comparable to the normal maize. While the favourable *opaque2* allele is recessive in nature, the endosperm modifiers that provide hard kernel texture (instead of the characteristic soft texture conferred by the *opaque2* allele in the absence of modifiers), are polygenic.

Introgression of *opaque2* along with endosperm modifiers into elite inbreds is complicated because of three major constraints: (i) each conventional BC generation needs to be selfed to identify the *opaque2* recessive gene, and a minimum of six such BC generations are required to recover satisfactory levels of recurrent parent genome; (ii) besides maintaining the homozygous *opaque2* gene, multiple endosperm modifiers must be selected; and (iii) rigorous biochemical tests are required to ensure enhanced lysine and tryptophan levels in the selected materials in each breeding generation, necessitating significant labour, time and resources.

Although conventional breeding procedures have been used successfully to convert commercial lines to QPM forms, the procedure is highly cumbersome and time-consuming. Scientists at CIMMYT (International Center for Wheat and Maize Improvement), Mexico have successfully developed an innovative combination based on SSR markers for *opaque2* allele and phenotypic selection for kernel vitreousness (through the accumulation of endosperm modifiers) for conversion of normal maize lines into QPM. Three SSR markers, umc1066, phi057 and phi112, which are internal repetitive sequences within the *opaque2* gene are being utilized in PCR-based assays to select individuals carrying the *opaque2* gene in successive backcross generations, and hence, the time required is reduced to one-half.

In addition, marker-aided background selection can help recover the same level of recurrent parent genome in three generations as would be achieved by six generations of conventional selections. Also, errors due to recombination frequency between the target gene and the flanking marker can be eliminated, since these SSR markers are located within the target gene itself. Apart from this, routine biochemical tests to ensure high lysine and tryptophan levels need not be carried out at each generation since the presence of *opaque2* is confirmed. Thus, MAS based on SSR markers for conversion of normal maize lines into QPM is simple, rapid, accurate, efficient, cost-effective and complementary to existing breeding protocols.

Marker-aided Pyramiding of Rice Genes

The most effective approach to combat bacterial blight caused by Xanthomonas oryzae pv. oryzae is the use of resistant varieties. So far, 25 resistance genes have been identifiedand some of them have been incorporated into modern rice varieties through conventional breeding. Transfer of Xa4 gene resulted in the development of many bacterial blight-resistant rice varieties, which have played a vital role in sustaining the rice yields in many countries. However, large-scale and long-term cultivation of varieties carrying Xa4 resulted in a significant shift of the dominant bacterial blight pathogens. Rice varieties with single Xa4 gene have become susceptible to the bacterial blight disease in many areas in Indonesia, India, China and Philippines.

One way to delay such a breakdown in bacterial blight resistance is to pyramid multiple resistance genes into rice varieties. This can be difficult or nearly impossible with conventional breeding approach due to epistasis of genes, particularly when a breeding line already carries a gene, e.g. Xa21 that shows resistance to most of the bacterial blight races at the time the varieties are being developed. With the conventional approach, a breeding line with Xa21 only cannot be distinguished from a breeding line with Xa21 plus some other genes. International Rice Research Institute (IRRI), Philippines and Punjab Agricultural University, Ludhiana have successfully employed MAS to pyramid genes for bacterial blight resistance.

All possible combinations of the four resistance genes, viz. Xa4, xa5, xa13 and Xa21 were pyramided using STS (sequence tagged site) markers. The pyramided lines show wider spectrum or higher level of resistance to the bacterial pathogen. Recently, the Government of Indonesia released two new rice varieties, 'Angke' and 'Conde' through MAS, which are resistant to bacterial leaf blight infection. Rice blast caused by the fungal pathogen, Magnaporthe grisea is another important disease throughout the world. The dynamic evolution of the virulence in the blast fungus coupled with the epistatic action of different resistance genes in the host plant, makes breeding for blast resistance a constant challenge. Molecular breeding approach is being widely employed for the improvement of blast resistance in many high-yielding commercial rice cultivars. Three genes, namely Pi1, Piz5 and Pita have been pyramided in a susceptible rice variety, CO39 using RFLP and PCR-based markers for durable blast resistance.

Most of the traits of agronomic importance are complex and regulated by several genes. Unlike the case of simply inherited traits that are controlled by one or a few major genes, improvement of polygenic traits through MAS is a complex endeavour. The difficulty in manipulating quantitative traits is related to their genetic complexity, mainly the number of genes involved in their expression and interactions among genes (epistasis). Because several genes are involved in expression of a quantitative trait, these genes, in general, have smaller individual effects on the phenotype, and the effect of the individual genes is not easily identifiable. This warrants repetitions of field tests to characterize accurately the effects of QTLs and to evaluate their stability across environments. The evaluation of QTL by environment interaction (Q × E) continues to be a major constraint on the efficiency of MAS.

Besides, epistatic interactions among different regions of a genome can induce a skewed evaluation of QTL effects. Also, if the genomic regions involved in the interactions are not incorporated in the selection scheme, they can potentially bias the selection process. Despite the proliferation of QTL mapping experiments in recent years, a number of constraints have imposed severe limitations on efficient utilization of QTL mapping information in plant breeding through MAS.

Salient among these constraints are:

i) identification of a limited number of major 'players' (QTLs) controlling specific traits;

ii) the notion that QTL identification is required whenever additional germplasm is used;

iii) inadequacies/ experimental deficiencies in QTL analysis leading to either overestimation or underestimation of the number and effects of QTLs;

iv) lack of universally valid QTLmarker associations applicable over different sets of breeding materials;

v) strong QTL - environment interaction; and

vi) difficulty in precisely evaluating epistatic effects. Increasing the efficiency of MAS for quantitative traits calls for improved field experimentations/ designs, robust mathematical models and sound statistical approaches. For instance, with composite interval mapping (CIM), field data from different environments can be integrated into a joint analysis to evaluate the Q × E interactions, thus, enabling identification of stable QTLs across environments. Besides, with a detailed linkage map, CIM allows a precise identification of the QTL in the genome and better identification of linked QTL (in coupling phase) from the same parental line. DNA-based markers also offer the unique advantage of identification and utilization of favourable QTL from even a phenotypically inferior parental line (in repulsion phase).

An especially powerful approach for enhancing the QTL mapping in tandem with MAS was proposed by Tanksley and Nelson. The advanced backcross-QTL (AB-QTL) analysis involves crosses between elite germplasm and unadapted genotype/wild relatives with favourable genes/QTLs, followed by two generations of backcrossing for developing several hundred sibling lines.

These lines, each retaining different genomic segments of the wild relative/unadapted genotype, are then genotyped using DNA markers. The lines become, in effect, a set of near-isogenic lines that individually dissect the effects of potential QTL in the background of the elite parent. At the same time, the BC lines also provide relatively fixed material for the critical step of replicated phenotypic evaluations. The essence of this strategy - uncovering and accessing the desirable alleles from wild relatives or unadapted genotype - demonstrates that QTL mapping can go hand in-hand with MAS rather than as sequential steps.

QTL-marker associations however, need to be unravelled through more intensive research efforts. It would be critical to find out whether extremely tight linkages between marker loci and QTL may lead to highly conserved allele associations. If so, observing marker allele frequency changes in long-term selection experiments or determining markers that explain significant portions of the combining ability variance in diallel or factorial crosses, might reveal universally applicable markers. Another approach would be to conduct QTL analyses in genetically broad-based panmictic populations by means of highly saturated integrated genetic marker maps.

For a maximal efficiency of MAS, direct QTL-allele-specific markers (such as STS markers derived from cloned QTL alleles) are needed. Although success in terms of cloning of QTL alleles is highly limited, map-based cloning and candidate gene approaches would increasingly facilitate isolation and characterization of agronomically important QTLs, particularly due to the rapid progress being made in genome sequencing of various plants.

Heterotic Performance in Maize

The maize lines B73 and Mo17 are not highly productive, but widely used across the world, and represent the two most important 'heterotic groups' in the US maize breeding programmes. Based on a mapping population derived from the B73 × Mo17 cross, QTLs contributing to heterosis for grain yield were mapped on nine of the ten maize chromosomes. Further, mapping studies suggested that two elite inbred lines, Tx303 and Oh43 contained genetic factors that might improve the heterotic response of the B73 × Mo17 single cross hybrid.

Six chromosomal segments each in Tx303 and Oh43 were transferred (using MAS) through three backcross generations into the target lines, B73 and Mo17. Two selfing generations followed the third backcross. Based on the initial evaluations of the testcross hybrids, the better-performing modified lines were selected for intercrossing and were designated as 'enhanced' lines. The single cross hybrids derived by crossing the enhanced B73 × enhanced Mo17 exceeded the hybrid 'checks'by 12 to 15%. This study was one of the earliest demonstrations that marker-facilitated backcrossing can be successfully employed to improve complex traits such as grain yield in maize.

AB-QTL Strategy in Tomato

The effectiveness of AB-QTL strategy was demonstrated in tomato by Tanksley and coworkers through a series of studies. Genetic enhancement, through AB-QTL strategy, for various traits of agronomic importance, including fruit quality and black mould resistance in tomato, were accomplished using wild relatives such as Lycopersicon pimpinellifolium, *L. peruvianum, L. hirsutum and L. cheesmanii.*

Effectiveness of MAS

Computer simulations have provided powerful tools for analysing the design and efficiency of MAS programmes. Three different selection strategies in a markerassisted background selection programme, namely twostep, three-step and four-step selections, were compared by computer simulation in terms of quicker recovery of a large proportion of the RPG. The simulations were based on maize genetic map with markers spaced about 20 cM apart and with the assumption that the target locus could be scored directly either through phenotype or a marker completely linked to the target gene. Major conclusions from this simulation experiment are as follows:

i) The four-stage sampling strategy which includes (a) selecting individuals carrying the target allele; (b) selecting individuals homozygous for recurrent parent genotype at loci flanking the target locus; (c) selecting the individuals homozygous for recurrent parent genotype at the remaining loci on the same chromosome as the target allele; and (d) selecting one individual that is homozygous for the recurrent parent genotype at most loci (across whole genome) among those that remain, is the most efficient procedure in general.

ii) With the four-stage sampling strategy and reasonable population size (50-100), one can expect to find BCprogeny with at least 96% RPG with 90% probability. It would take six generations of traditional backcrossing to reach this stage, besides the risk of a larger probability of linkage drag around the target gene.

iii) Increasing the number of markers genotyped at each generation had little effect. Once the threshold of one marker per 20 cM is reached, additional markers (except perhaps around the target locus) would not be required. The frequency of recombination, and not the number of markers, is a more important limiting factor in reducing linkage drag, which suggests that sampling larger populations with fewer markers makes sense than the reverse.

Developments in MAS Strategies and Genotyping Techniques

Single Large-scale MAS

In this strategy, selection of parental lines is first carried out among outstanding elite material for the trait to be improved with the best allelic complementarity. By crossing each selected parental line with a tester (elite line lacking the target trait), segregating populations are developed. Genomic regions of interest for each parental line are identified by combining favourable alleles in the segregating populations (e.g. Ffamilies and RILs). MAS, based on reliable PCR-based markers to fix favourable alleles at target genomic regions, is conducted only once on large segregating populations derived from crosses between the elite lines. The strategy is suggested to offer three distinct advantages:

i) favourable alleles (selected to improve a specific trait) are derived from two or more sources of elite parental lines in a complementary scheme, disregarding the 'recipient/donor' line concept;

ii) plants with fixed favourable alleles at specific genomic regions are selected at an early generation of recombination and no pressure of selection is applied outside the targetted regions. This assures good allelic variability in the rest of the genome for future line development under various conditions and environments; and

iii) the approach is particularly relevant for pyramiding favourable alleles at cloned genes or major QTLs in new germplasm.

Pedigree MAS

This approachis especially relevant for crops such as wheat, where pedigrees of elite germplasm are known. Fingerprinting elite wheat materials must be conducted in a set of lines actively used in the breeding programme, and in elite materials to be subsequently released. The data may be combined with the phenotypic data collected during different selection cycles to identify favourable alleles for trait(s) of interest. For example, if an elite line contains alleles for yield performance in a target environment, their frequency should be higher than the expected random frequency in offspring derived from this elite parental line.

This shift in allelic frequency reflects phenotypic selection by breeders and may be identified by comparing fingerprinting data of both parents and their offspring. Once the favourable alleles are identified, DNA markers closely linked to the target genomic regions can be used to accelerate fixation of favourable alleles in the next selection step: a new set of elite materials to derive the next set of elite lines. This MAS strategy was suggested to be most efficient when conducted on For Fsegregating populations.

Breeding by Design

'Breeding by design' is a novel concept that aims to control all allelic variations for all genes of agronomic importance. The authors of this conceptpropose that by understanding the genetic basis of all agronomically important characters and the allelic variation at those loci, the breeder would be able to design superior genotypes in silico, which demonstrates that DNA markers not only improve selection processes but can aid in creating novel genotypes bearing new characteristics of agronomic importance.

The strategy involves three well-defined steps: (i) mapping loci involved in all agronomically relevant traits, preferably through introgression line (IL) libraries that are extremely powerful not only in mapping, but also in reducing the complexity of polygenic traits by separating them into a set of monogenic loci. An IL library consists of a series of lines harbouring a single homozygous donor segment introgressed into a uniform cultivated background. (ii) Assessment of the allelic variation at those loci through intensive chromosome haplotyping and extensive phenotyping of all agronomic traits. Haplotypes refer to a combination of linked marker alleles that occur in a locus in a set of accessions. (iii) Designing superior genotypes comprising a combination of favourable alleles at all loci through accurate selection of recombination events using flanking markers to collate different favourable alleles next to each other. 'Breeding by design' involves an integrative, complementary application of technological tools and materials currently available to develop superior varieties. However, success of this approach would essentially depend upon availability of extremely saturated marker maps and precision of phenotyping.

PCR-amplification

The efficiency of MAS in larger breeding populations could be improved if the laboratory throughput is high and costs are low. Currently, the most common way to evaluate the outcome of PCR amplification is to analyse the amplified products using gel electrophoresis once thermal cycling is complete. This process takes extra time and effort and also increases the chance of laboratory equipment or reagents being contaminated. Consequently, attempts have been made to develop means to detect the accumulation of PCR products either during thermal cycling or following thermal cycling without having to remove the samples for processing.

With the advent of third-generation markers, such as SNPs, dispensing with electrophoresis as a core technology to detect polymorphism has become the primary focus. All major SNP detection techniques rely on an initial PCR amplification of the target DNA segment, while they differ in the ways in which the alternate amplicon sequences are discriminated from one another. The SNP is targetted either internally within the amplicon or/at immediately downstream of the 3′ end of one of the amplification primers, using allele specific oligonucleotides. The probes employed in both molecular beacon and technologies are based on the principle of fluorescence resonance energy transfer.

Recently, non-electrophoresis-based PCR assays for the allelic discrimination at two linked loci flanking an important QTL controlling days to pollen shed in maize were described. The assays are based on the fluorogenic 5′-nuclease procedure (Taqman), which allows for direct detection of the PCR product by the release of a fluorescent reporter dye as a result of DNA amplification (Figure 1). The main steps in the reaction sequence are polymerization, strand displacement and cleavage. Two dyes, a fluorescent reporter (*R*) and a quencher (*Q*) are attached to the fluorogenic probe. When both dyes are attached to the probe, reporter dye emission is quenched.

During each extension cycle, the DNA polymerase cleaves the reporter dye from the probe. Once separated from the quencher, the reporter dye fluoresces. The assays were developed after sequencing the alleles at both the loci, by designing suitable primers and probes based on SNPs or insertion/deletion polymorphisms. The Taqman procedure allowed for a fast and highly reproducible analysis directly in the PCR vials. The complete automation of the process makes this technique highly valuable for the largescale screening required for MAS studies and map-based cloning projects.

Similarly, high throughput genotyping using Taqman probes for a polymorphism linked to SCN resistance gene, *Rhg4* has been standardized. This method can be used for scoring polymorphism in an RIL population and for scoring parental lines in breeding programmes. The Taqman method of determining genotype was accurate in 90% of scores in the RIL population compared to 95% accuracy with electrophoresis. Therefore, this method can be applied to automated large-scale genotyping for soybean breeding programmes.

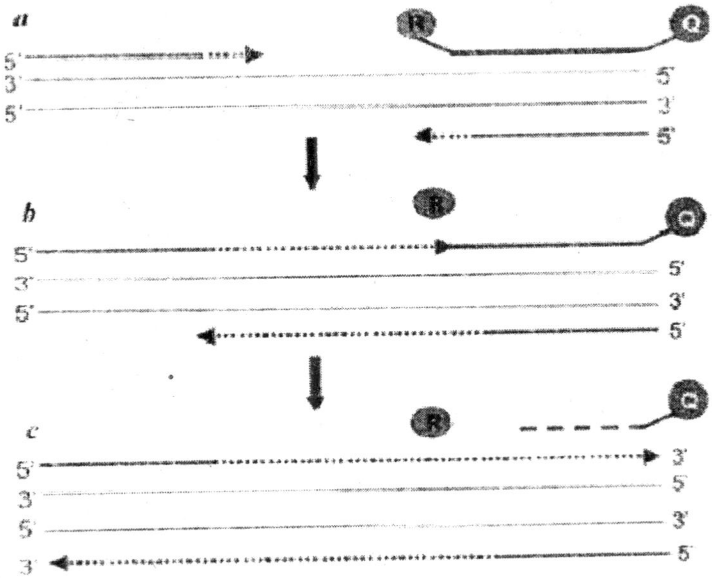

Figure 1. Schematic diagram of Taqman assay. a, Polymerization; b, Strand displacement and cleavage; c, Polymerization completed. R, Reporter; Q, Quencher.

The molecular beacons have been demonstrated to be useful in discriminating between transgenic and non-transgenic transgenic cereals. The principle of molecular beacon assay is explained in Figure 2. The hairpin stem formed by the complementary arm sequences of the beacon cannot coexist with the double helix that is formed when the probe hybridizes to its target. Consequently, the molecular beacon undergoes a conformational change that forces the arm sequences apart and causes the fluorophore to move away from the quencher, thereby allowing fluorescence.

Oligonucleotide ligation assayand dynamic allelespecific hybridizationhave been developed, which can potentially increase the power of DNA genotyping.

Another interesting and innovative assay is invader SNP assay, which is PCR- and gel-free. Invasive cleavage is a novel non-PCR technique relying on the action of a special endonuclease to remove overlapping oligonucleotides during polymerase-induced extension. These assays are performed on total genomic DNA using universal invader and fluorescent probes and an allele-specific signal probe.

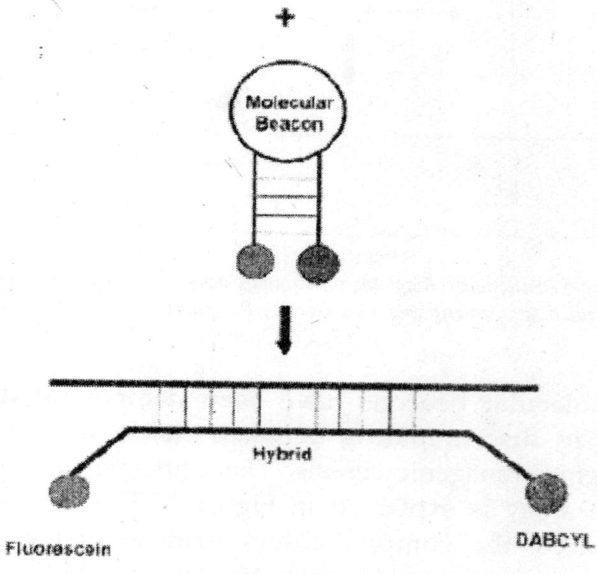

Figure 2. Molecular beacon assay.

The invader assay employs endonucleases (derived from Archaea species) from the FEN family (cleavases) to cleave structure-specific rather than sequence-specific sites. Allele-specific oligonucleotide probes (ASOP) are designed to produce the structure necessary for cleavage of the signal probe. If ASOP matches the sequence of the allele, then the required flap structure is produced and the signal probe is cleaved. If the ASOP mismatches the sequence of the allele, then the required flap structure is not produced and signal probe is not cleaved.

The simple semi-automated data interpretation of the assay allows the screening of 25,000 to 50,000 genotypes per operator per week, which is tenfold greater than gel-based SNP and microsatellite typing. The SNP typing in this assay is found to be highly accurate when compared to standard PCR-based methods. The assay requires no special instruments other than a microplate reader and proves to be a highly sensitive method for detecting SNPs directly from the genomic DNA with high genotyping throughput.

Integrating molecular marker technologies such as MAS into breeding strategies could become increasingly important in the coming years, to realize genetic gains with greater speed and precision. Although MAS is currently used more widely for simply inherited traits than for polygenic traits, with the development and access to reliable PCRbased markers such as SSRs and SNPs, in several crop plants, efficiency of genotyping large populations or breeding materials has significantly increased. Marker-assisted foreground selection is already gaining rapid momentum as allele-specific markers are now becoming available, through crop genomics research, on a number of agronomically important traits.

The promise of MAS for improving polygenic traits in a quick time-frame and in a cost-effective manner is still elusive. There is a wider appreciation that simply demonstrating that a complex trait can be dissected into QTLs and mapped to approximate genomic locations using DNA markers would not serve the ultimate goal of trait improvement. As suggested by Young, research on quantitative traits need to utilize better scoring methods, larger population sizes, multiple replications and environments, appropriate quantitative genetic analysis, various genetic backgrounds and, whenever possible, independent verification through advanced generations or parallel populations. In facing the challenge of improving several lines for quantitative traits, MAS strategies should probably concentrate on using DNA markers in one key selection step to maximize their impact. They could be used at the very beginning of the scheme as predictive tools to reduce the number of crosses, at an early stage of recombination to fix target genomic regions, or at an advanced stage of germplasm development as a diagnostic tool, when the allelic value has already been identified.

The field of 'functional genomics' is making rapid strides, aiming towards assessing the gene functions through genome-wide experimental approaches. Innovative tools such as DNA chips, microarrays, expressed sequence tags, etc. help in the quantitative assessment of RNA levels in the sample, which in turn might benefit the breeder in selecting best lines based on RNA expression profilesas much as marker genotypes.

These innovative approaches are not only expected to identify more genes involved in regulating different pathways in response to stress conditions, but also mark a significant paradigm shift in the field of molecular breeding. In future, quantitative genetics will look towards genomics for information to develop more biologically meaningful models, while genomics will look towards quantitative genetics to develop and validate hypotheses involving complex gene interactions.

Bioinformatics will play an important role in facilitating this crossover. Thus, integrating genomics and bioinformatics into the field of molecular breeding is expected to bring in more fundamental revolutions in plant breeding. It would be crucial to recognize the complementarities between molecular technologies and conventional breeding and harness this vital synergy into developing comprehensive research strategies aimed towards more efficient crop improvement in future.

Bibliography

Fukuda, W.M.G.; Oliveira, S. de; Iglesias, S.; Iglesias, C., "Cassava breeding", *Crop Breeding and Applied Biotechnology,* 2002.

Heisey, P.W., M.A. Lantican and H.J. Dubin, "Assessing the benefits of international wheat breeding research in the developing world", *The Global Wheat Impacts Study,* CIMMYT, Mexico, DF, 1966-1997.

______________., C.S. Srinivasan and C. Thirtle, "Public sectoı plant breeding in a privatizing world", *Agricultural Information Bulletin* No. 772. Economic Research Service, USDepartment of Agriculture, Washington, DC, 2001.

Henson-Apollonio, V., "Patent protection for plants: Exceptions to infringement", *InWIPO-UPOV symposium on the co-existence of patents and plant breeders' rights in the promotion of biotechnological developments,* Geneva, UPOV, 25 October, 2002.

Mooney, P.R., "The law of the seed: another development and plant genetic resources", *Development Dialogue,* 1983.

Moore, G. and W. Tymoski, "Guide to the International Treaty on Plant Genetic Resources for Food and Agriculture", *The World Conservation Union (IUCN),* Gland, Switzerland, 2005.

Morris, M.L. (ed.), *Maize seed industries in developing countries,* Lynne RiennerPublishers, Boulder, Colorado, 1998.

______________., "Impacts of international maize breeding research in developing countries",1966-98. CIMMYT, Mexico, DF, 2002.

Pickett, A.A. and N.W. Galwey, "A further evaluation of hybrid wheat", *Plant Varieties and Seeds,* 1997.

Plucknett, D.L. et al, "Crop germplasm and developing countries", *Science,* 1983.

Sperling, L., J. Ashby, E. Weltzien, M. Smith and S. McGuire, "A framework foranalyzing participatory breeding projects and results", Euphytica, 2001.

Welch, R.M., "Breeding strategies for biofortified staple plant foods to reduce micronutrient malnutrition globally", *Journal of Nutrition,* 2002.

_______________.; Graham, R.D., "Breeding crops for enhanced micronutrient content", *Plant and Soil,* 2002.

Index